E-What?

A Guide to the
Quirks of New Media
Style and Usage

PUBLICATIONS FROM EEI PRESS

Error-Free Writing: A Lifetime Guide to Flawless Business Writing

The Great Grammar Challenge: Test Yourself on Punctuation, Usage, Grammar—and More

Letter Perfect: A Guide to Practical Proofreading

Mark My Words: Instruction and Practice in Proofreading

My Big Sourcebook: For People Who Work with Words or Pictures

The New York Public Library Writer's Guide to Style and Usage

Real-World Newsletters To Meet Your Unreal Demands

Stet Again! More Tricks of the Trade for Publications People

Substance & Style: Instruction and Practice in Copyediting

The Editorial Eye: Focusing on Publications Standards, Practices, and Trends (subscription newsletter for editors, writers, managers, educators, journalists, and anyone who cares about language)

E-What?

A Guide to the Quirks of New Media Style and Usage

BY THE EDITORS OF EEI PRESS

EEI PRESS®
A Division of EEI Communications®

E-What? © 2000 by EEI Press, reprinted July 2003. All rights reserved. Beyond copying a few pages for personal use, no part of this book may be duplicated, reproduced, or stored in any form or in any medium without written permission from the publisher. Reviewers may quote brief passages in critical articles and reviews.

EEI Press publishes other books and *The Editorial Eye*, a subscription newsletter focusing on standards and practices for excellence in publications. Visit the Web site, www.eeicommunications.com/press, to view sample articles and book excerpts.

Call 800-683-8380 to inquire about quantity orders. Discounts are offered to bookstores, corporations, nonprofit associations, universities and colleges, conference resellers, and training organizations. For more information, call or write to us at the address below.

EEI PRESS®
A Division of EEI Communications®

EEI Press | A division of EEI Communications
66 Canal Center Plaza, Suite 200
Alexandria, VA 22314-5507
tel 703-683-0683; 800-683-8380 to order
e-mail press@eeicommunications.com

Library of Congress Cataloging-in-Publication Data

E-What? : a guide to the quirks of new media style and usage : how to handle inconsistencies in punctuation, capitalization, Internet addresses, and more / from the editors of EEI Press.
    p. cm.
  ISBN 0-935012-24-9
  1. Telecommunication--dictionaries. 2. Computers--Dictionaries. 3. Authorship--Style manuals.

TK5102 .E94 2000
808'.066004--dc21

                                                                                     00-063632

This book was produced with QuarkXPress 4.1 on a Macintosh G3, using Minion, Agenda, and Zapf Dingbat fonts. It was printed by Signature Book Printing of Gaithersburg, Maryland.

Contents

Foreword vii

Acknowledgments xi

1 Keeping Up with Style 1
    Abbreviations, Acronyms, and Initialisms 3
    Capitalization 7
    Wild-Eyed at Uncultivated Words? 12
    Prefixes 15
    Suffixes 17
    Compounding 17
    Morphing Parts of Speech 19
    Symbols 20
    Numbers 20
    Web Addresses 21
    Writing for the Web 24
    Writing about the Web 32
    Writing E-Mail 34
    Working with Word 36

2 Quick Reference List of Troublesome Terms 37
    Rules, Advice, Your Best Guess 37
    Why Are These Particular Terms Here? 38
    Glossary 40

3 Planning Your Online Style Guide 87
    What's Our Style? 88
    Creating a Guide 90
    Making Intranet Content Easy to Use 100
    Making It Work 105

Annotated Bibliography 111
   Electronic Sources 111
   Glossaries, Thesauruses, and Acronym Lists Online 113
   E-Mail Discussion Lists 114
   Printed Books 115

Sidebars
   Consider Creating a Style FAQ 13
   Language Under the Influence of New Media 15
   Confusing Domain Name Categories 22
   Getting a Feel for Address Formats 23
   Metric Isn't Dead Just Because We Don't Use It 25
   New Media, as She Is Written About 27
   Taking Disabilities into Account 89
   Two Kinds of Dialogue with Online Style Guide Users 90
   Case Study: Using a Macro to Check for
      Biased Language 92
   Case Study: Using a Macro to Convert Tables 94
   Intelligent Links 101
   Reality: They're Going to Print Out Pages 104
   A Web Site Needs a Style Guide, Too 108

Foreword

I UNDERSTAND THAT SETTLING STYLE AND GRAMMAR RULES can be difficult when you're working on Internet time. That's why the good people at EEI Communications have been studying these issues; have a look at their work at eeicommunications.com.

Er, I mean, www.eeicommunications.com.

No, make that http://www.eeicommunications.com. On the World Wide Web, that is.

To seasoned online pros, any of those three formations would be enough guidance to find the Web home of a company. Given the number of people to whom this is all still new, however, it's not good enough to assume that all readers grok the same Internet dialect. (Though we can often pick up clues from context—you got *grok*, didn't you?)

The Internet would be easy to understand and talk and write about if only it would stop changing every week—but instead, novel terms of art turn into hackneyed clichés at about the same frequency that dot-com startups rip up their business plans. In my experience, certainty in matters of style doesn't outlive the first e-mail from a reader asking, "How do I get to that site? The address you gave didn't work." I have answered perhaps a thousand of these messages since mid-1994, when I started writing about the online world for *The Washington Post*. In a lot of ways, I'm still figuring out what is and is not obvious about all this.

The linguistic target keeps moving, and I don't expect this situation to change for a long time. The point of the text we're working on shouldn't keep moving around, however, and that's why a guidebook like this is helpful. In self-defense, as we try to keep up with the fluidity of Internet style, we need some baselines.

Here's a baseline we can all agree on: Writing and speaking about the Web, and about any other technical topic—quantum physics, the history of NASA's Apollo program, automotive maintenance, or Bruce Springsteen bootlegs—should aim to nail down meanings without clouting the thumb of the everyday reader. Unnecessary jargon, unexplained acronyms, and jarring inconsistencies can quash attention and slow comprehension.

Consider, for instance, the issue of how to describe Usenet newsgroups—a topic that, five years ago, provoked constant arguments on the alt.internet.media-coverage newsgroup (itself a casualty of Internet evolution—here today, gone today). Coverage in the mass media would most often describe these forums as *Internet message boards* or *online bulletin boards.* The true believers, who had been holding forth on Usenet for years before any newspaper had an inkling that a story could originate there, flamed (that is, vituperatively criticized) the "old media" for "not getting it." Meanwhile, the few journalists online at that point argued, essentially, "Look, nobody's going to be that confused by this term, and at least people will have a chance of getting the analogy."

Those journalists had a point, but they were wrong. Not referring to Usenet as Usenet or newsgroups as newsgroups meant that readers had no easy way of trying this resource out for themselves with their own software, which didn't and still doesn't use terms like *message boards* or *bulletin boards.* Details do matter—especially since readers can go check things out for themselves.

A similar debate has been going on about how to specify a Web site's location (which, incidentally, I don't call *a URL* unless I'm writing for a technical audience—what's wrong with "address"?). If you simply specify the domain name, you can save a lot of trouble all around. One, plenty of Web-based companies include the *.com* extension in their names, so the name *is* the Web address. Two, almost all current Web browsers will fill in the *www.* prefix for you—why make the reader type in four extra characters? But on the other hand, there's no shortage of Web addresses that *don't* start with *www.*—not to mention the few heretical Web addresses that depart from the standard *http://.* (For instance, some bank sites start with *https://,* with that *s* denoting an encrypted connection secured against eavesdropping.)

Hence, I continue to insist on the full Web address. I am losing this fight, though, both at my office and in this book. It grates on me—why not be consistent and minimize reader confusion?—but I have, so far, been unable to drag everybody else along. (People become quite passionate about these matters. We editors take pride in the inventive patches we've used to construct our preferred-usage crazy quilts.)

I do have one consolation, though: In another year or so, the fashion for addresses, along with other Internet-related conventions, may have changed yet again. But we're not going to make deadlines, and we're not going to be taken seriously, if we start from scratch stylistically every day. We can't wait until matters of style settle down; we have work to do today.

This book can help you settle an argument or start one, if you need to, perhaps with the pedant down the hall who insists on lowercasing *the web* but wants you to know he is a *Webmaster*. Editors are still the arbiters of much that is happening in publishing; they have their work cut out for them. Fortunately, they also have *E-What?*

Rob Pegoraro (rob@twp.com)
Consumer electronics editor
The Washington Post
Spring 2000

Acknowledgments

THE TWO PEOPLE WHO RESEARCHED AND WROTE MOST OF *E-What?* are Kathryn Hall, director of electronic information for Computer Economics, Inc., in Carlsbad, CA, and Lee Mickle, an EEI Communications editor in Alexandria, VA. Hall has also project-managed production for countless print publications; Mickle has imposed style on hundreds of print, Web, and intranet publications.

The people at EEI Communications who helped develop and fine-tune the content are Robin Cormier, Keith Ivey, Linda Jorgensen, Claire Kincaid, and Candee Wilson. Scott Baur, Sue Flint, Alison Foster, Ed Gloninger, Sherrel Hissong, Rebecca Hunter, Catherine Malo, Merideth Menken, Patricia Moran, Kristina Rutledge, Judi Sigler, and Christine Stevenson copyedited, designed, and produced the book.

Special thanks to the stylish outside reviewers—each with enough experience in writing, editing, publishing, and computing to disagree heartily about e-style—who offered valuable criticism: Gabriel Goldberg, Rob Pegoraro, and Phil Smith III.

Finally, thanks to the many *Editorial Eye* subscribers who asked us to write this book, and to the dcpubs, copyediting-l, and wiw e-mail discussion list members who answered our call for topics and terms that cause them most editorial pain. We hope *E-What?* helps them survive in a world that insists on creating products with new-media-influenced names like *IncrEdibles*—microwavable foods (one is eggs!) in a push-up tube. Aside from the fact that the instant association is with "inedibles," the name is asking to be misspelled. But it's also asking for attention, and it wins on that count.

Samuel Johnson said, "Language is the dress of thought." If so, then style is the cut of the cloth. Sometimes style fits effortlessly, as if cut on the bias. Lately its skimpy seams have been pinching a lot of nerves. The publisher hopes *E-What?* will help communicators as they hold the fort against editorial weirdnesses and find rationales for styling inescapable (and—admit it—refreshing) innovations. Here's to a workable truce!

1. Keeping Up with Style

BY JANUARY 1, 2000, A MENTION OF THE Y2K BUG ALREADY carried a hint of nostalgia, and as you read these words, it seems as quaint and faintly annoying as *23 skiddoo*. Yet soon after the term *dot-com* was coined, it was accepted as useful shorthand to indicate a company that does business chiefly or entirely over the Internet. It also might show up in any of three forms: *dot-com, dotcom,* or *dot.com.*

What's going on here? Shall we blame the many-splendored media for making us so easily jaded? Is our tolerance for new language really changing to reflect a higher speed limit for daily life? Are consistent spelling, style, and usage inevitable casualties of that tolerance?

Speed. The information revolution is essentially a change in the speed of communication—and communication, in turn, affects just about every field of human endeavor. It's a commonplace to say that language evolves; but the term evolution implies change so gradual that it can be observed only in retrospect.

Now, though, language is changing so fast, coming and going, that the most conscientious can't keep up with the variations. First-ever, one-of-a-kind terms are coined hourly—just put an *e* in front of any word, and it's understood to mean "something that takes place electronically, most likely through the Internet." And *i* isn't far behind, neck and neck with *cyber*. The result is countless neologisms like the ones listed here:

- Eccentric capitalization (lowercased initial capital letters, upper-cased midcaps, a mixture of both, and special characters) is a technique vendors use to name what they hope will be perceived as the latest and greatest and cutting-edgiest of technologies. (For example, *1stUp.com* and *eBusiness*.)
- Hybrid constructions abound—should it be *cyber cash* or *cybercash*?
- Abbreviated forms may be quirky but obvious and literal (*RLSI*: Ridiculously Large-Scale Integration), while others are illogical and arbitrary (*WNIC*: Wide-Area Network Interface Co-Processor).
- Back-formations (nouns used as verbs—for example, as in *E-mail me*) take root, and most of the people who use them care nothing about following traditional grammar rules or formal style.

And some things are created purely to shock, show off, seem cutting-edge. We can't print off-color examples here; this is a family style guide. But here's a tame example: *By e-enabling our services we will grow revenues in all of our functions.* Wow, *e-enabling!* (See the Quick Reference List of Troublesome Terms on page 37 for many other examples.)

Is there a way to make sense of this linguistic hash? There must be—otherwise, we'll simply be engaged in *failing* to communicate, at unprecedented speed. In fact, language and our understanding of language are astoundingly flexible. Even though we may not be able to cite chapter and verse from an English textbook, native speakers apply "the rules" of grammar and word formation to these new terms almost without thinking.

It's when we stop to think that we start getting confused, like the fabled caterpillar who could walk just fine until someone asked him how he kept track of all those legs. Yet we *have* to stop and think about terms for which we have no precedent. How are they functioning? What's the nearest analogue? We *have* to find a way to see patterns of similarity that will allow us to style unfamiliar terms consistently for our readers.

In the midst of so much change, it's nice to have some reassurance. That's one of the functions of reference works—to back up our gut feelings with authority, or, on rarer occasions than we might suppose, to set our feet back on the correct path. Sometimes we encourage you to trust yourself to discern and follow the consensus, as we present it here and as you daily absorb newly emerging bits of evidence. In other instances, we offer linguistically based guidelines that you can apply when the rules haven't caught up with reality.

There is no reason to throw out perfectly good rules for clear writing merely because it will be read on a screen, so in yet other instances we invoke the same guidelines you've heard before.

No special postmodern attitude is required in order to balance speedy innovation (a cultural inevitability) with consistency that shows concern for reader comprehension (professionally admirable—and a good business move). The trick, of course, is knowing which is which is which is which—and when to make up your own damned rules. (See Planning Your Online Style Guide, page 87.)

We take a consensus approach to usage and style. That means we derive the rationales for our decision making from a number of authoritative contemporary sources. One of them is us. EEI Press editors wrote a major style guide (*The New York Public Library Writer's Guide to Style and Usage,* for HarperCollins) and each month publish an award-winning subscription newsletter, *The Editorial Eye.* We've critiqued and edited and written many hundreds of client publications.

For this book, we browsed far and deep: consumer, trade, technical, and scholarly periodicals, as well as major newspapers. We take aim at what people tell us are their biggest sources of uncertainty and frustration—without aspiring to produce another computing style guide. *E-What?* is meant to be mainstream. But we cannot and do not pretend to provide a rule for every occasion, or to map all the occasions that may and will arise. This is a primer.

We can show a healthy respect for the time and intelligence of our readers when we make thoughtful decisions about style and usage. As *Read Me First! A Style Guide for the Computer Industry* reminds us, "Consistency is not just some abstract goal to be achieved for its own sake; rather, the intention is to reduce the impact of the mechanics of communication on readers."

So even though there will not always be a perfectly consistent resolution for every usage quirk in the major problem categories that follow, please remember that it *is* the thoughtful attempt that counts.

ABBREVIATIONS, ACRONYMS, AND INITIALISMS

Abbreviations, the generic term, are meant to be a time- and space-saving form of shorthand, but they travel with a fair amount of stylistic baggage.

Two main types of abbreviations are used to represent a longer name or phrase:

- Acronyms are formed from the first letter or letters of a group of words. Acronyms can be pronounced as a word (e.g., *ASCII* is said as AS-key).
- Initialisms are also formed from the first letter or letters of a group of words. But initialisms are pronounced letter by letter (e.g., *IBM* is said as I-B-M).

The more common if inexact term seems to be *acronym,* so we'll use it in the following discussions to mean initialisms as well.

To Define or Not to Define?
That's one of the first questions that arise with regard to acronyms. Here, as in so many areas, you must apply some judgment. One person's alphabet soup is another person's everyday speech. Would you rather be considered condescending or arrogant? Sometimes it seems those are the choices. In an article for an employee newsletter, it would probably be silly to spell out the name of every department, if everyone in the company routinely refers to them as HR, S&H, or the like. A national press release from the same company should err on the side of caution by assuming that some readers will be stumped by such acronyms.

The high-tech world is rapidly catching up with the military as the number-one coiner of acronyms. Some, such as *modem,* have already undergone what's called *acronymy*—the process by which abbreviations come to be perceived and used as words. If you referred to a component that converts digital signals to sounds and back again as a *modulator/demodulator,* few people would know what you were talking about. (*Modem,* by the way, is an example of a syllabic acronym—formed from the first syllables of the basic term, rather than initial letters. More recent examples include *pixel,* short for picture element, and *codec,* short for coder-decoder.)

WYSIWYG, on the other hand, is listed in *Merriam-Webster's Collegiate Dictionary,* tenth edition, but still written in all capitals—a sign that its assimilation is not yet complete. Today, a computer magazine needn't bother to define it; some general-interest publications perhaps still should. In a couple of years, though, defining it may be as unnecessary as defining IRS, FBI, or COB.

What to do? "Know your readers" is the catchphrase in deciding which acronyms to define on first use. Here's another guideline: When in doubt, spell it out. Even if 90 percent of your readers know what URL stands for, the other 10 percent will be grateful to find out without having to stop and look it up. Of course, the newer an acronym is, the more important it is to define it, unless your audience is a crowd known to be Web-conversant and would see overdefinition as editorial amateurism. (This is one of many times you'll have to make an educated guess about such matters.)

Which brings us to the point of this guide: Is there a difference in how you use acronyms when you're writing for the Web? The answer is yes. The Writing for the Web section (page 24) describes some of the physical differences involved in reading material in print and onscreen. These differences imply that you should use acronyms sparingly. Your readers may not have time to familiarize themselves with acronyms as they quickly browse, may not encounter the place where they are defined, and may not follow the jump to a definition link. An even more basic reason: Clusters of capital letters detract from readability.

When you encounter a new acronym, how do you find out what it means? General-purpose acronym dictionaries run to multiple volumes, are outdated before they're printed, and are not very useful for defining short acronyms, since anything less than four letters is apt to have a couple hundred possible definitions. It's better to use a subject-specific reference. Three handy online references for definitions of computer-related acronyms and terms are www.geocities.com/ikind_babel/babel/babel.html, www.acronymfinder.com, and www.netdictionary.com/html/index.html.

Using Articles with Acronyms

In general, you can simplify your life by leaving off articles (*a, an, the*) with acronyms. For example, *IEEE publishes about a fourth of the world's electrical and computer engineering papers* (where *IEEE* is the Institute of Electronics and Electrical Engineers). If you do use them, you must be aware not only of the acronym's meaning, but of whether it's an acronym or an initialism—whether people familiar with the term pronounce it as a word or a string of letters. For example, the acronym SME (subject matter expert) is common in technical writing. Do you use *an* (as in *an S-M-E*) or *a* (as in *a smee*)? Turns out it's pronounced *smee*. URL also takes *a*—

it's pronounced U-R-L. The conundrum arises with any acronym that starts with F, H, I, L, M, N, R, S, U, or X.

Capitalizing Definitions

In general, acronyms (formed from the first letters of all or only the main words in a phrase or name) are written in all caps, without periods. Note, however, that when you define an acronym, the words themselves are not capitalized unless they represent a proper name. Compare *WYSIWYG* (what you see is what you get) with *ASCII* (American Standard Code for Information Interchange). Sometimes it can be tricky to ascertain whether the term is considered proper; you may see *HTML* (Hypertext Markup Language) spelled out with or without initial caps. But it's a protocol name, and it should be uppercased.

Obviously, the consensus tack doesn't work for arbitrary, proprietary terms. You just need to have a good computer dictionary handy (see a list of useful such references in the Annotated Bibliography, page 111). Having a friend in your local IT department wouldn't hurt. And you'd definitely benefit from creating a personal stylesheet (or online style guide) to help you manage the specialized terms you're most likely to encounter.

As noted, an acronym may eventually be accepted as a word, in which case it sheds its capitals. But some acronyms that aspire to the status of words may find their place already taken—IT (information technology) is unlikely to be written *it,* no matter how familiar the term becomes.

Overlapping Acronyms

It's sloppy practice and potentially confusing to use a single acronym to represent more than one term. If you define IS as "information services," don't use it to mean "information systems" as well. Spell out the term that's less used or less commonly represented by an acronym.

Singular and Plural Acronyms

Most acronyms are singular in meaning. You form the plural just as you would for a word—by adding a lowercase *s*. One *CD-ROM,* two *CD-ROMs.* A few, however, are defined as plural, such as IS for information systems. There's no need to add an "s" to such acronyms—but you do need to be careful not to use them as singular terms (*an IS*). Be kind to your readers—be consistent!

If you're making a plural form of an initialism that isn't written in all caps, use an apostrophe for the sake of clarity:

Learn your *abc's*.

Mind your *p's and q's*.

It isn't necessary to add an apostrophe to the plural form of syllabic acronyms:

modems

pixels

Abbreviations

Abbreviations range from the commonplace (Mr., Mrs.) to the esoteric (Kbps). As these examples demonstrate, it's harder to generalize about the capitalization and punctuation of abbreviations than about acronyms and initialisms. You must be precise, especially when the abbreviation represents a unit of measure. For example, *kb* stands for kilobit, *kbps* for kilobits per second; *kB* stands for kilobyte, *kBps* for kilobytes per second. (All are measures of speed, just as *mph* is.)

Like other neologisms, abbreviations are proliferating. Many are informal curtailments of words that go from being used in conversation to being seen in print, such as *app* for application and *sig* for signature.

As for defining them, the same guidelines as for acronyms apply. Define any abbreviations that are likely to be unfamiliar to most of your readers. In a technical text laden with abbreviations and acronyms, it may make sense to provide a glossary in addition to or instead of in-text definitions. And remember, a glossary doesn't have to be the last item the reader sees. If it's essential to an understanding of the text, put it up front. In a document published on the Internet, you can link each abbreviation to its definition.

CAPITALIZATION

It's easy to say "capitalize proper nouns" and "treat brand names the way the owners treat them." But trying to follow these rules is far from simple. Is LISTSERV a proper noun? (The answer is yes.) If so, what is its generic equivalent—assuming it has one? (Electronic mailing list.) If a brand name begins with a lowercase letter, what happens when it's the first word in a sentence? Is there an authoritative source for the spelling and capitalization of brand names and trademarks?

What's Proper?

It's the *Internet,* also called *the Net.* It's the *World Wide Web,* frequently called *the Web.* These are proper nouns and should hang onto their capitals for dear life.

When you use *Web* as a modifier, retain the capital. A location on the Web is a *Web site* or *Web page.*

When you use *web* as a prefix, lowercase it. A person (male or female) who maintains a Web site is a *webmaster.* A magazine published exclusively on the Web is a *webzine* (also called an *e-zine*).

On the other hand, your company, my company, and the company on the floor below us can each have its own *intranet,* so the term is not capitalized.

Business Names and Trademarks

Capitalization of business names, trademarks, and service marks should follow the owner's preference. But sometimes that's tricky to ascertain. In the good old days, you were fairly safe using headline-style capitalization (capitalizing only the first of each "important" word). Now, lots of manufacturers are using midcaps (capitalizing seemingly random letters inside a name instead of only the first letter or significant letters), as well as eccentric punctuation, to distinguish product names. If Kleenex were being marketed today as a piece of software, it would probably be written *klee*NeX!*

It's not safe to take your guidance solely from other publications such as computer magazines that review new products. You have no way of knowing how carefully they've done their research. Perhaps the best resource for questions of spelling and capitalization of new products is on the Net itself—specifically, on the owner's site. Announcements of new products should appear there with the names accurately reproduced. In addition, you'll find "how to contact us" information, so you can follow up by e-mail or phone if necessary.

Many sites have a convenient Legal, Copyright, or Trademarks page. Find it through the site navigational structure (usually in a "corporate information" section) or site search engine. A few worth mentioning:

- Microsoft: www.microsoft.com/trademarks/t-mark/g-lines.htm
- Hewlett-Packard: www.hp.com/abouthp/trademark.html

- Canon: www.usa.canon.com/copyright.html
- Adobe: www.adobe.com/misc/copyright.html

An owner's site should represent spelling and capitalization faithfully, but some actually *don't* represent their own trademarks correctly. For example, the E*Trade Web site always spells its name E*TRADE. Try e-mailing the public relations, marketing, or legal department for an authoritative answer. (An interesting solution suggested by one editor is to look for a picture of the software box on the company Web site. A picture that's clear and big enough to read can verify the spelling. But packaging itself may be the victim of design enhancements that disregard strict typographical rendering.)

When referring to your own trademarked products, use the ® symbol for a registered trademark, the ™ symbol for an unregistered trademark, and the ℠ symbol for a service mark. Use them each time you mention the products; the use of the symbols reinforces your claims as rights holder.

But if you're not the mark holder, you're under no legal obligation to use those symbols. Just spell and capitalize the names of products and print them as nearly like the original marks as you reasonably can. Since owners have enlisted novel capitalization and punctuation to distinguish their products, however, you're going to be faced with some bizarre-looking constructions.

What if you want to begin a sentence with *eBay* or end it with *Yahoo!*? (Note the need for a question mark after the exclamation point.) The first choice is to try to avoid the issue by reworking the sentence. If that fails, however, retain the name's idiosyncrasies. And if you can't stand the way the resulting sentence looks, go back to square one—now you have even more incentive to figure out a way to reword the sentence:

Instead of: eBay is a popular online auction service.

Try: One popular online auction service is eBay.

Instead of: Do you use the search engine Yahoo!?

Try: Is Yahoo! one of the search engines you use?

Yahoo! itself uses the tagline "Do you Yahoo!?" It's an infinitely concentric set of silliness we get ourselves into with such trademarked end-punctuation games. It's enough to bring on an attack of *e-ennui*.

Here's a short list of trade names for quick reference, but be aware that there are thousands of others, with more being coined every day:

3Com	Compaq iPAQ
AccuBooks 2000	CompuServe
Adobe Acrobat	Corel CATALYST
Adobe ActiveShare	Corel KnockOut
Adobe After Effects	Corel NetPerfect
Adobe FrameMaker	Corel PHOTO-PAINT
Adobe FrameViewer	Corel Print House
Adobe GoLive	Corel VENTURA
Adobe ImageReady	Corel WordPerfect
Adobe ImageStyler	CorelDRAW
Adobe InCopy	CorelXARA
Adobe InDesign	Corex CardScan
Adobe LiveMotion	DirecPC
Adobe PageMaker Plus	DIRECTV
Adobe PageMill	E*Trade
Adobe PhotoDeluxe	eBay
Adobe Photoshop	eMachines' eMonster,
Adobe PostScript	eTower, eSlate, and eView
Adobe PressReady	ePregnancy.com
Adobe Streamline	Evite.com
AGFA SnapScan	Family Tree Maker
Aldus FreeHand	flooz
Allaire ColdFusion	GeoCities
AltaVista	GO.com
Amazon.com	HomeGain.com
America Online	HotBot
Ameritrade	Hotmail
AOLTV	HotSync
Ask Jeeves	HP DeskJet
Ask.com	HP PhotoSmart
Autodesk AutoCAD LT 2000	HP ScanJet
BizRate.com	HP SureStore drives
Canon PowerShot	iBaby.com
CD-ROM	IBM ThinkPad
Compaq Deskpro	iBook

iMac
Intel Celeron processor
Intuit Quicken
Intuit TurboTax
Iomega Jaz drive
Iomega Zip drive
iPlanet.com
iVillage.com
iWon.com
JavaScript
JobsOnline.com
Kodak PalmPix
Kodak PhotoDoc scanner
LabelWriter
LaGarde StoreFront
LendingTree.com
LEXIS-NEXIS
Linux
LookSmart
Macromedia Authorware
  Attain
Macromedia Dreamweaver
Macromedia Fireworks
Macromedia Shockwave
MapQuest.com
MarketWatch.com
McAfee ActiveShield
McAfee VirusScan
Metacrawler
Microsoft BackOffice
  Small Business Server
Microsoft FrontPage 2000/
  PhotoDraw 2000 V.2
Microsoft IntelliMouse
  with IntelliEye
Microsoft MSN
  HomeAdvisor
Microsoft Visio 2000

MP3
Napster
NEC SuperScript printer
NETGEAR
NetMAX FileServer,
  WebServer, FireWall
NetObjects Fusion
Netscape Navigator
Nikon Coolpix
Nikon Coolscan
Norton AntiVirus
Optiquest
Paint Shop
Palm Pilot
Palm Vx
PayMyBills.com
PayPal
Peachtree Accounting
PocketMail
Polaroid SprintScan
PostScript
priceline.com
QuarkXPress
QuarkXTensions
QuickBooks
QuickCam Pro
Quicken 2000
QuickTime
Rand McNally StreetFinder
Rand McNally TripMaker
RealNetworks
RealPlayer
Red Hat Linux
RoboHELP
Rocket eBook
SageMaker
Sony PictureBook
Sony SuperSlim Pro

SpeedStep (Intel)
Stamps.com
Symantec WinFax PRO
The Motley Fool
TrueType
uBid.com
UMAX PowerLook scanner
USPS eBillPay
ViewSonic
VisualCafé
WebCrawler
WebMail
WebMD.com
Windows 95
Windows 98
Windows 2000
Windows NT
Xerox DocuPrint
Yahoo!

Personal Titles

This is another area in which the old rules are being bent. The trend toward downstyle means lowercasing personal titles, even when they precede a person's name:

chief information officer Wang Lee (abbreviation: CIO)
content provider Rolf Stuart
sysadmin Adrian Lowry (short for system administrator)
webmaster Keith Ivey

WILD-EYED AT UNCULTIVATED WORDS?

"They're coining words all over the place! There's no logic, no discipline, no order!" Yep, the barbarians are at the gates—again. Or to be more accurate, they've already swum the moat, climbed the walls, entered the citadel, and proceeded to spread hardy weeds bound to overtake our cherished cultivars. Is that just about enough invasion metaphors?

We don't have to start writing or, as readers, tolerating gibberish just because the boundaries of language are elastic. Extremes are still extremes, even when the extreme has become, so to speak, the norm. We can introduce the least appetizing of new terms to civilization. But we shouldn't assume that all new words are coined senselessly or perversely. As proof, here's how Tim Berners-Lee, in *Weaving the Web,* describes the thoughtful way he chose the name for his creation: "Friends at CERN gave me a hard time, saying it would never take off—especially since it yielded an acronym that was nine syllables long when spoken. Nevertheless, I decided to forge ahead. I would call my system the 'World Wide Web.'"

Visit a Web site like the Jargon File (www.tuxedo.org/~esr/jargon/) or pick up a book like *The New Hacker's Dictionary*, which is based on the Jargon File, and you will find thousands of entries—both new coinages and new definitions for existing terms. We live, as the Chinese curse says, in interesting times. Of course, order is important in writing; we need to agree on the meaning of terms to keep from tripping up our readers. We need to be consistent in our treatment of terms for the same reason. We also need to stay sane in the process.

What does sanity mean, in this context? Perhaps most of all it means keeping in mind that the coinage of new words is a healthy part of linguistic evolution. When people discover a new world or invent a new technology, the creation of new words accelerates to keep up with events and experience. Since new media technology is communication technology, not only does it spawn new terms, it also distributes them at lightning speed. And then discards and replaces them just as fast.

The pace of creating language matched to technological advances sometimes precludes reasoned consideration of how to treat a new form—by the time you've figured out whether to capitalize it, how to hyphenate it, or whether to even allow it, you may be describing a technology on its way out. Sometimes you just have to go with the flow, even if it feels more like whitewater kayaking than rafting down the Mississippi.

CONSIDER CREATING A STYLE FAQ

The University of Chicago Press FAQ offers hyperlinks to questions and answers of particular interest to those of us trying to keep up with new-media style vagaries. The address is **www.press.uchicago.edu/Misc/Chicago/cmosfaq.html**.

You might consider compiling an essential FAQ for the persistent issues that keep people guessing in your own organization. Whether you place such a FAQ on your intranet or hand out a printed version to new employees with other orientation materials, you'll be enabling people to make consistent style decisions. That's something you can do right now to improve the quality of your publications.

Keeping Up with Style

Making Style Choices
When you do have a chance to make a style decision, welcome it! It's an opportunity to create order—or at least, a tidal pool—from a sea of variations. Here are a couple of points to bear in mind:

- Your choice should reflect the existing personal or corporate style you have been using up until now—if you have been. If you have an "up" style (a tendency to capitalize words) and prefer to hyphenate prefixes and unit modifiers, you may choose to write E-mail rather than e-mail or email.

- The tendency in English (to which there are many exceptions, naturally) is for terms to go from two words (cyber cafe) to a hyphenated term (cyber-cafe) to one word (cybercafe). Where you decide to come down on this continuum is likewise a factor of your existing style, as well as how familiar the term is likely to be to your readers. It's also a good idea to look around and see how the rest of the world is treating the word (although you won't find complete agreement, or you wouldn't have to be making a decision in the first place).

We've listed our preferences for a number of these terms in the Quick Reference List of Troublesome Terms. We provide patterns on which you can base some decisions about new terms as they pop up in the publications you work on. And we'll need to update our decisions in a year or so, just as you will. What to do when no reference has a paradigm or rule to offer?

If you're trying to style a term your own organization or profession has added to the lexicon—for which you can find no precedent—ask the smartest copyeditor you know (the one with a penchant for wearing shades of gray) what makes most sense. Or send your wayward term to *The Editorial Eye,* eye@eeicommunications.com, for help. We'll catalog it for a future edition, and we'll give you advice in the meantime.

Dealing with Geekspeak
The folks who toil in the trenches writing code or developing new hardware don't mind being called geeks; it's a badge of honor. Nor do they mind speaking a jargon that serves to identify members of the clan and bewilder outsiders—that's one of the functions of jargon. When you

encounter it, pause for a moment. Can you provide enough context to guide your readers to its meaning without eliminating its flavor? If you can, go ahead.

Resources can help you decipher geekspeak. If you were writing about cowhands, sailors, or truckers, you wouldn't purge their speech of anything that wasn't standard English. Rather, you would use it to give a vivid picture of their lives. Although they'd probably laugh to hear it, there's a similar aura of romance and mystery—even superstardom—around today's geeks. There's a nice bit of justice in that.

PREFIXES

High-tech communications are spawning prefixed terms as fast as they're spawning acronyms. The evolutionary trend in English is to go from hyphenating prefixes to writing them closed up, or solid. The question is, how quickly does the process occur? The speed of communications is spilling over into the speed of linguistic change—some writers figure that since the hyphen will disappear eventually, they might as well dispense with it at the outset.

Thus, you see ebusiness, email, etrade, and even eeconomics—as one ad says, eenough! In the case of single-letter prefixes, we recommend retaining the hyphen rather than rushing the process. For one thing, a single-letter prefix tends to look strange without a hyphen—sort of like pig latin. It upsets spell checkers. And not all of these terms will stick around; a shakeout is likely.

LANGUAGE UNDER THE INFLUENCE OF NEW MEDIA

Even in the world of low-tech print publications like newsletters, we're all guilty of LUI—language under the influence of new-media terms. At least one free electronic newsletter has replaced the traditional *premium* offered to new print subscribers with an *e-freemium*. It still means a little something extra offered as part of a marketing pitch, but adding the *e-* makes the company sound much more "wired." The linguistic variety that arises from the innovation and energy associated with new-media possibilities may drive some copyeditors crazy, but it's a sign of a vital culture. Until we become bored hearing *e-this* and *e-that*, that is.

cyber- short for cybernetic. Generally closed up. *cybercafe, cybercash™, cyberpunk* (a subgenre of science fiction; the term was coined to describe William Gibson's seminal *Neuromancer,* published in 1984), *cybersex, cyberspace* (from *Neuromancer,* describing the "consensual hallucination" that is approximated by the World Wide Web), *cybersquatting* (registering domain names early in hopes of selling them back to companies later on). However, if a coinage seems likely to be a one-time use, you may prefer not to close it up: "Welcome to Cyber Purgatory."

digi- short for digital. Like *e-* and *cyber-,* this prefix is used loosely, often to give a "hip" look to certain words. *Digibabble, digirati,* and *digitocracy* are soundalike coinages whose meanings are easy to infer.

e- short for electronic. Implies "done over the Internet." Keep the hyphen in generic terms: *e-banking, e-business, e-commerce, e-mail, e-text.* Also frequently used in product names: *eBay, Rocket eBook™* (NuvoMedia), *ePaper®* (Adobe).

hyper- *hyperlink, hypertext.* Refers to the system that enables readers to jump from one document to another on the Internet. The dictionary definition of *hyper* is "excessive," and the use of links may verge on the excessive. (See Writing for the Web, page 24.)

i- short for Internet. Same implication as *e-*. Slightly less common than *e-,* but used the same way. It may be used as the prefix for a Web site name (*iParenting, iVillage*).

meta- short for more highly organized or specialized. *Meta-analysis, metadata, metatag.*

multi- short for multiple. *Multitasking, multithreaded.*

techno- short for technical. *Technobabble, technophile, technophobe.*

tele- short for distance. *Telephone* and *television* remind us of how quickly we assimilate new words formed with prefixes. Younger, but already almost as familiar, are such terms as *telecommute* and *teleconference.*

web- short for the World Wide Web. You have to be careful with this one, since sometimes it's a noun used as an adjective: *Web page, Web site, Web-related.* When *web* is used as a prefix, it isn't capitalized: *webcam, webmaster, webzine.*

SUFFIXES

Novel terms formed with suffixes aren't nearly so prevalent as those formed with prefixes. Perhaps, once we've tacked an e- on the front of every word in the language, we'll start relying more on suffixes.

-cast refers to a means of disseminating information: *broadcast, narrowcast, netcast, webcast.*

-tech from *technical: high-tech* (also seen as *hi-tech*), *low-tech.*

-ware from *software* and *hardware,* we go to *freeware, groupware, shareware,* and *wetware* (the brain). The plural *warez* (pronounced "wares") is used specifically to refer to computer software and hardware. Ware will we go next?

COMPOUNDING

Put any two words together and use them to describe a third word, and you have a compound modifier. Put them together and use them to represent "a person, place, or thing," and you have a compound noun. The question is, how do you put them together? Do you join them with a hyphen? Do you keep them pristinely apart when they appear at the end of a sentence, but insert a hyphen when they precede the word they modify? Do you throw caution to the wind and weld them together immediately?

Trends and Patterns

The speed of electronic communications is accelerating certain trends in grammar. One of these trends is the tendency of terms to evolve from two words, to a hyphenated term, to one word. Some terms start and stay hyphenated. Some skip step one and maybe step two and go immediately to either a hyphenated form or a single word formed from two words. *Voicemail* is good example. Some people are still writing it as two words; still others have decided that it's one word. A few used a hyphen, but not many, and not for very long. Below are some terms that we've seen used variously:

	anti-aliasing	antialiasing
double click	double-click	doubleclick
home page	home-page	homepage
news group		newsgroup
user group		usergroup
voice mail		voicemail
Web site, web site	Web-site, web-site	Website, website

Another trend is to discard certain so-called rules that are difficult to apply. One of these is the rule for hyphenating unit modifiers depending on their position in a sentence. Traditionally, such modifiers are hyphenated only if they precede the word they modify—unless the hyphenated form is used so frequently that it is considered a permanent compound. Here's an example:

I like to invest in *high-tech stocks.*

Some of these *companies* are so *high tech* that I don't understand what they do.

Today, however, many writers are impatient with this rule; they feel that if a compound term is likely to become permanent, it should always be hyphenated. When speed is of the essence, they reason, who has time to fiddle with guessing where finicky hyphens are on the chart?

Here are some terms that we nominate to the status of permanent compounds:

double-click (also left-click, right-click)
high-tech
ill-(anything)
leading-edge
off-the-shelf
password-protected
user-friendly
Web-based
well-(anything)

If : then

When writers coin a new word, one way to decide how to treat it is to see what existing word(s) it resembles. Thus, we can justify the following coinages:

database → knowledgebase	front page → home page
desktop → laptop, palmtop	postmaster → webmaster

MORPHING PARTS OF SPEECH

Yet another characteristic of our language is our ability to form verbs from nouns and vice versa. And—guess what?—such mutations are happening faster than ever. Is *e-mail* a noun or a verb? In all but the most formal writing, it's both. *I'll e-mail my reply. I just received your e-mail.*

A *cookie* is an identifier that Web marketers can use to identify computers that visit their sites. Overheard in a conversation, the phrase "We will cookie you" passed without the listener exhibiting visible tremors of discomfort. But the speaker probably would not have written it the same way—or would he? So if you encounter a word being used as a part of speech that isn't assigned to it in the dictionary, it may not be wrong; it may simply be new. Comfort levels differ; when this sort of sporting spoken language is asked to sit still in print, it loses some of its verve. A little of it goes a long way in all but the most genuinely, strenuously geeky of publications (for example, *shift* magazine).

A Partial List of Nouns That Can Be Verbs (Or Is It the Other Way Around?)

The words below have essentially the same meanings in their noun and verb forms. Some are relatively new; some have been around quite a while. The list does not include words such as *port* that have different meanings as nouns and verbs (see that entry in the Quick Reference List of Troublesome Terms).

access	fax	mail	spam
address	firewall	modem	spec
cache	FTP	OEM	Telnet
chat	hit	post	undo
chunk	host	program	upload
download	index	search	visit
e-mail			

Keeping Up with Style

SYMBOLS

When working with electronic communications, you may see or hear the names of some unfamiliar symbols:

angle brackets ‹ › (You may be accustomed to thinking of these as the mathematical symbols for "less than" and "greater than.") Used to enclose HTML tags.

asterisk * Of course, you know this one, but you may not be used to seeing it used for *emphasis* in place of bold or italic face. Because e-mail programs don't offer a choice of fonts, people resort to using symbols. People may also type words in all caps for emphasis, but this has come to be considered rude— the visual equivalent of shouting.

at sign @ Used in e-mail addresses between the user alias and the name of the host computer. Not used in URLs. Also commonly used in advertising copy for a trendy look (or so it is hoped).

forward slash / or backward (back) slash \ A forward slash is used to separate parts of a URL (www.eeicommunications.com/press/) and a back slash is used for computer file names (C:\Documents\draft.doc). It's important to use the right one.

tilde ~ Sometimes part of a URL, as in the Jargon File URL: www.tuxedo.org/~esr/jargon/. May also be used in computer tags or codes because it is rarely found in ordinary English text.

underscore _ An underscore character _before_ and after a word may be used in place of italics, as the asterisk is also used, when e-mail or a text-only system does not permit the use of regular italics.

NUMBERS

There's no difference between "Internet style" and print style when it comes to numbers. If you're writing on technical or scientific topics, you're more likely to use figures for numbers above nine and for all units of measure. In general, whatever style you normally follow for print publications will work. The subject of information technology itself can entail discussion of very large numbers, which is where certain prefixes come in:

kilo- (often k)	1 thousand
mega-	1 million
giga-	1 billion
tera-	1 trillion

WEB ADDRESSES

A Web site address is usually called a URL: uniform (not universal) resource locator, generally pronounced U-R-L, although some people pronounce it "earl." URLs follow conventions, just as postal addresses do. Whereas postal addresses go from the specific (individual's name) to the general (city and state), URLs go from the general (Internet protocol name) to the specific (host name and miscellaneous details about specific pages). Because URLs are read by literal-minded machines, it's crucial to use internal punctuation accurately.

Here's a typical URL: **http://www.eeicommunications.com/eye/utw/.** Here's how it parses:

Protocol. Here, it is **http.** Most, but not all, URLs start with *http,* which stands for Hypertext Transfer Protocol. Browsers today are programmed to assume that it's there, which means you don't actually have to type it. Another protocol is *FTP,* which stands for File Transfer Protocol; if you omit the introductory address tag *ftp,* you won't reach your destination. (*Gopher* and *Telnet* are two other protocols, but you'll rarely encounter them.) In a URL, the protocol is always followed by a colon.

Host Name. Whatever falls between the double slash and the first single slash (or the end of the address) is the host name. The host is the computer server that maintains the file. Most (but not all) host names start with **www.,** which stands for World Wide Web. The host name here is **www.eeicommunications.com.** The **.com** (pronounced "dot-com") specifies the type of organization. Currently, URLs classify organizations according to these conventions:

.com	commercial	**.mil**	military
.edu	educational	**.net**	network
.gov	governmental	**.org**	organizational
.int	international		

But this is not a strict system; many dot-coms automatically register the .com, .net, and .org domains to prevent misuse of their hoped-for

Keeping Up with Style

CONFUSING DOMAIN NAME CATEGORIES

Keith Ivey, the EEI Communications webmaster, says that the rules for some of the domain name hierarchies have broken down in recent years because of market forces. Companies that are not service providers can and do use the **.net** designation instead of the **.com** (commercial) extension. In fact, InterNIC (the organization responsible for the domain name hierarchy) gave up trying to enforce restrictions on **.org** (for nonprofits) and **.net** years ago. Essentially, **.org** and especially **.net** are now just alternatives for organizations when their first-choice **.com** name is already taken. In fact, Network Solutions' registration process encourages people to register **.net** and **.org** at the same time when registering a **.com** domain. It's all about money and controlling brand identity.

eventually huge name recognition. Most people have heard about www.whitehouse.com, the porn site that gets bombarded by schoolchildren trying to find www.whitehouse.gov. As Net use expands, more specific organizational types may have to be added. Host names are not case-sensitive, but are usually written in lowercase letters.

Miscellaneous details. The host name will be enough to get you to an organization's home page. Elements following the first single slash will take you to a specific page within the site. Here, **/eye/utw** specifies the home page for "Untangling the Web," a set of articles with the **eye** (short for *Editorial Eye* newsletter) section of the site. This part of the URL can be quite long and, unlike the host name, *is* case-sensitive.

International Sites

The six classifications above apply chiefly to sites in the United States. Most non-U.S. URLs end in a two-letter country abbreviation. Some of these are obvious: *uk* is United Kingdom; *fr* is France. Others are harder to guess: *de* is Germany (Deutschland); *ch* is Switzerland (Confoederatio Helvetica, the Latin name). You can find a list of about 200 non-U.S. URLs at www.geocities.com/ikind_babel/babel/babel.html.

Typesetting URLs

How do you set a URL in type? In general, URLs are not case-sensitive, so take the easy way and type them in lowercase letters. For brevity, omit *http://* and start with *www*. It's preferable to set URLs on one line of type, without breaking them, but that's not always possible. When you must break a URL, do so after a slash but before a period, to avoid the appearance of terminal punctuation. That is,

> *www.eeicommunications.com.*

Never insert a hyphen when breaking a URL; since internal punctuation is part of the address, adding a hyphen where it doesn't belong can misdirect readers.

GETTING A FEEL FOR ADDRESS FORMATS

Understanding the structure of URLs will help you spot errors. Here's one mistake that appeared in a newsletter: **www/will-harris.com**.

In a Web address, the part between the double slash and the first single slash (or the end of the address) must contain at least one period and must end in a valid top-level domain. Knowing that, you might guess that the correct address is this: **www.will-harris.com**.

When a sentence ends with a URL, it's pretty safe to use terminal punctuation. Even if readers misunderstand and assume that the period is part of the URL, almost all browsers will ignore the end punctuation and take them to the right place.

How about setting off URLs typographically? It's a question of style, not rock-ribbed right or wrong. Some publications set them in italics or in bold italics. This has the advantage of making it easy to see where the URL begins and ends, although in practice, that's usually pretty obvious. We recommend following older conventions: If you don't use a special type for a postal address or telephone number, why do it for a URL?

Microsoft Word automatically underlines anything it recognizes as a URL or an e-mail address, showing that the program has created a link to that address. In a Word file, you can double-click on such links and go from Word to the address in question. This fact has led many people to typeset URLs with an underscore. But that's an option, not a necessity. See "Working with Word," page 36, for information about changing Word's settings.

E-Mail Addresses
These are shorter than URLs and can usually be recognized by the use of @. Here too, punctuation is important. Some systems use periods in the addressee's name:

    linda.j.smith@bigtown.gov

Others go with first initial, last name:

    lsmith@bigtown.gov

Again, e-mail addresses aren't case-sensitive, but people may be. Some folks just feel more polite using capitals for names:

    Linda.J.Smith@bigtown.gov

If you prefer to send e-mail this way, don't hesitate. But when citing Smith's e-mail address, type it lowercase, to avoid implying that the capitals are essential.

WRITING FOR THE WEB
Is the Web a different world? Must we discard all we know and learn to write in a whole new way? Are we free to break the shackles of grammar and style, to express ourselves however we choose?

Ahem! Well, yes and no. Writing for the Web is different from writing for a prind publication, but the difference is comparable to that between, say, writing for a daily newspaper and writing for a reference book. The factors that matter are *who* the readers are and *how* they will interact with the medium. Writers have been taking those factors into consideration since they first set quill to papyrus.

What Is Web Style?
Good Web style is tailored to the needs of readers. Does this statement sound familiar? Of course. One thing the Web makes far easier than print communication is knowing just who "the readers" are. No matter how carefully books are marketed, they can wind up just about anywhere. Subscription publications use an array of demographic surveys, polls, feedback, and ZIP code analysis to figure out who their readers are and what they want.

The Web gives publishers an awesome (and some feel, frightening) amount of insight into just who visits a particular site, how long they spend viewing it, and whether they return. Writers for the Web also really can know who their readers are. They have a corresponding responsibility to tailor their writing as closely as possible to their audience. There should be at least as much difference between Web sites for various audiences as there is between, say, *Seventeen* and *Modern Maturity*.

But you can pick up either of those magazines and find it written in (largely) comprehensible English. Sentence structure is the same in both. Each adheres to a consistent style with regard to such matters as capitalization, punctuation, and the use of numbers versus words. The same holds true of writing for the Web. English is English, not Webbish. A few conventions are different, but they quickly become second nature. Magazine readers understand the use of jump lines (for example, *continued on page 46*); Web readers take links in stride.

In the narrow sense of the word "style," there's no reason to deviate from what you use in print. If you follow AP style in your print publications, follow it in your Web publications. There is no universal Web style, just as there is no universal print style. But there are conventions accepted by first-rate Web publishers:

- Writing for the Web should be concise. A widely accepted statistic is that people read 25 percent slower on a computer screen than on paper. The corollary is that a document written for the Web should be 25 percent shorter than one written for print.

- Writing for the Web should use short paragraphs, lots of subheads, and displayed lists. These guidelines are dictated in part

> **METRIC ISN'T DEAD JUST BECAUSE WE DON'T USE IT**
>
> When writing for an international audience, remember that most of the world uses the metric system. When providing U.S. measurements for distance and speed and the like, also include (alongside, in parentheses) the metric equivalent. (Check your dictionary for a conversion chart.) Also, avoid using the # symbol to indicate *pounds*; a single quote to indicate *feet*; and double quotes to indicate *inches*. These marks aren't recognized everywhere outside the United States.

by the computer's limitations. Despite advances in display technology, it's still not as comfortable to read a computer screen as a printed page. It's also difficult to distinguish certain punctuation marks, such as commas and semicolons, on screen. And many special characters don't display at all unless they are carefully coded. For more information about the legibility of text onscreen, see Planning Your Online Style Guide, page 87.

Writing for the Web should be strong enough to capture the online reader's attention. A print magazine reader can flip at will to perhaps 20 articles; on the Web, literally millions of sites are mere clicks away. Not only brevity but fresh, lively writing and accurate delivery of information are requisites for keeping readers. These criteria matter more than ever in print because of the sheer range and quantity of Web publications, many free; these criteria matter more than ever online for the same reason.

Tone

A bit of conventional wisdom is that Web writing is more personal and less formal than print writing. This is one of those grand statements that breaks down a bit upon examination. Good writing is personal wherever it appears, in the sense that reading is a personal experience—one reader, one document. The best writing seems to speak directly to the individual reader, even if, in fact, those readers number in the millions.

As for "less formal," it's true that the second person is widely accepted and that new terminology (whether slang, jargon, or just the latest catchphrase) is prevalent. It's not true that errors of spelling or grammar that would embarrass the publishers of a printed document are acceptable on the Web. In fact, since readers of Web pages may have little information about the publisher, such errors detract seriously from a document's credibility: "If these people can't even check their spelling, how likely are they to check their facts?"

How Readers See You

Writing for the Web differs from writing for print chiefly in how readers approach it. They may not go straight to your site, but may reach it from a link at another site, or as a result of entering a topic in a search engine. You cannot assume that they will start reading at the beginning of your document and continue until they reach the end. They see only what can be displayed on their computer screen—perhaps the equivalent of one-third of a printed page. If they're sufficiently interested in what they see, they may scroll up or down, click on internal links, or print the entire document. Or, they may simply glance at the screen to see if it contains the information they were searching for and, if it doesn't, move to another site.

This situation has consequences for both writing and design. Writers need to think in terms of screens' worth of material—usually about 1,000 words per screen—unbeautifully referred to as "chunks." Each chunk should be somewhat self-contained, with links connecting it to the other chunks of the document. If you're accustomed to outlining before you write, you may want to experiment instead with diagramming, sketching how each chunk of your document will relate to the others.

Design

The design of Web sites is the subject of numerous books. A Web site, like any publication, will benefit from the services of a skilled graphic artist who understands how to make a site readable and visually appealing. Writers can and do design their own sites, and there is software that makes it reasonably easy, just as desktop publishing software makes it possible to design your own print publications. Still, a professional designer will do a professional job.

Beware if someone tells you that what you want "can't be done" on the Web. That person is speaking from ignorance or laziness. The choice of fonts, typefaces, sizes, styles, and special characters is as wide as in

NEW MEDIA, AS SHE IS WRITTEN ABOUT

Kay Tallis, a network programming manager and documentation trainer, advises that, when writing about technology and computing—especially for a high-tech audience—it's best to avoid anthropomorphism. That is, don't describe computers or programs in terms of things they "want" a user to do or "won't let" a user do. Program commands don't "take" or "send" users anywhere; users give commands to gain access to programs. Tallis says that when computers hear us talking about them in a condescending, personalized way, "It tees them off."

print—not to mention effects that aren't available in print, such as animation and sound. On the other hand, you get nothing for nothing; expect to pay for top-notch, original design. If you can't afford to go that route, expect to spend quite a bit of time learning about Web design and coding. Same old song—what you pay for in print publications, you pay for in electronic publications.

Graphics

In the early days of the Web, designers tended to rely on graphics to attract and hold readers' attention. But the novelty of graphics for their own sake has long since worn off. A well-designed Web site should certainly incorporate *relevant* graphics—but gimmicks are no substitute for strong content, and strong content requires good writing.

Links

Technology can enable readers to jump from one document to another, but it can't (yet!) change the way the mind works. People tend to process material in a linear fashion; we can't really "read" more than one document at a time. A reader who jumps away from your document may never return. So restrain your use of links to those that are genuinely valuable.

You can easily incorporate useful links into your document if you think of them as cross-references. The only trick is to phrase links as part of the text. Instead of writing "See the section on Chicago for more information," write "One study took place in Chicago during 1998." Note that this is a link to another section of *your* document. You can also link to other sites, but do so with caution. Readers who leave your site may go off on a treasure hunt, jumping from link to link, and never return to your document. It's safer to group external links at the end of the document or at the foot of the Web page, where they're less distracting to readers.

Abbreviations

If your writing contains numerous abbreviations, acronyms, or initialisms, you may want to take a second look. Because readers see only a screenful of information at a time and may or may not read your entire document, the print technique of defining such terms on first mention won't always work. Consider spelling out terms you might ordinarily abbreviate, if you think readers won't know them. If it's essential to use abbreviations, create links to their explanations.

Writing for the World
The World Wide Web is exactly that. Unless you're writing for a subscription-only site, your work is available for viewing around the globe. Can you afford to be blind to this fact? Increasingly, the answer is no. The notion of having readers in Iceland, Serbia, Indonesia, and Ghana can be daunting—and should be inspiring, too.

Technical writers, in particular, are compelled to be aware of such issues as use of idioms that cannot be translated literally, cultural references that make no sense out of context, and even the use of certain colors in illustrations. Some companies have developed rigid vocabularies, with as many don'ts as do's, to facilitate computer-assisted translation of technical documents. People who complain that such vocabularies are limiting miss the point. The vocabularies are not intended as tools of literature; they are designed to facilitate basic communication that will enable customers to use a product.

Even if you aren't constrained by the knowledge that your work will be translated into 25 different languages, you can apply some commonsense global sensitivity to your writing for the Web. Here are a few points to bear in mind.

Currency. If you're referring to U.S. dollars, say so. ("The book costs US $30.75.") Here's a list of countries that call their currency the dollar: Australia, the Bahamas, Barbados, Belize, Brunei, Canada, the Cayman Islands, Dominica, Fiji, Grenada, Guyana, Hong Kong, Jamaica, Kiribati, Liberia, Namibia, Nauru, New Zealand, St. Kitts-Nevis, St. Lucia, St. Vincent and the Grenadines, Singapore, the Solomon Islands, Taiwan, Trinidad and Tobago, Tuvalu, the United States, and Zimbabwe. There's plenty of overlap in other currencies too (e.g., dinar, franc, peso). Use words rather than symbols, as well as the name of the country, if there is any likelihood of confusion. Using words also eliminates having to search for special symbols on your computer. If you're writing about a particular time, you might also want to add the year (1998 U.S. dollars). Because currencies fluctuate constantly, it's not advisable to give equivalents in other currencies unless you're writing about the past.

Clock times. Particularly if you know you're communicating with an international (or even a nationwide) audience, it's a good idea to specify a time zone if accuracy is crucial. And here again, spell it out. "The meeting will begin at 2:00 p.m., Eastern Standard Time." Technical or military readers, as well as many international audiences, are comfortable with the 24-hour clock: "The meeting will begin at 14:30, Eastern Standard Time." If you decide to go with this system, mention it, to avoid confusing some readers.

Dates. Abbreviations can lead to confusion. 2/6/01 may be February 6 to you, but in European countries such as France and Germany, the date normally precedes the month, so readers may perceive it as June 2. For maximum clarity, write "February 6, 2001." If your writing is so full of dates that you feel abbreviations are essential, give readers a brief explanation: "In this document, the day appears first, then the month, then the year. For example, 2/6/01 is June 2, 2001."

Of course, the above discussion isn't global either—it assumes a Christian (and, specifically, Western, as opposed to Eastern Orthodox) calendar. January 1, 2000, didn't signal a new millennium (or the close of an old one, to be painfully precise) in India, where the year was 1921; in Israel, where it was 5749; or in Muslim countries, where it was 1420. However, unless you expect a sizable portion of your readership to be in one of those areas, it's safe to use the Western calendar; most readers will understand it.

Numbers. There are two issues to consider: punctuation and numbers above one million. Where the United States uses commas to punctuate figures larger than four or five digits, Germany and France, for example, use periods, and Sweden uses spaces. The period/comma dichotomy is more likely to lead to misunderstandings, since a period could be mistaken for a decimal point. If your writing contains just a few large numbers, spell them out to avoid confusion. Otherwise, give an explanation: "In this document, 1,000 equals one thousand; 1.00 equals one."

For numbers above one million, the arithmetic values actually diverge. For example, 1,000,000,000,000 is a trillion in the United States but a billion in Great Britain. *Merriam-Webster's Collegiate Dictionary*, tenth edition, gives a cogent explanation in its table of numbers. For scientific or technical writing, it's a good idea to use figures for such large numbers. If you spell them out, specify *(U.S. system)* after the words.

Punctuation. American-style punctuation in English differs from British punctuation. There's some disagreement as to whether you should adopt the latter for Web publications, assuming you don't use it for your print materials. If you can reasonably expect or are actively seeking an international audience, and particularly if your subject matter is technical, it's worth considering.

American: Double quotation marks for a primary quotation; single quotation marks for a quotation within a quotation. Terminal punctuation marks within quotation marks:

> After exclaiming, "This so-called 'disaster' is all just a misunderstanding!," Roberts turned on his heel and stormed out.

British: Single quotation marks for a primary quotation; double quotation marks for a quotation within a quotation. Terminal punctuation marks inside or outside quotation marks, depending on sense:

> After exclaiming, 'This so-called "disaster" is all just a misunderstanding!', Roberts turned on his heel and stormed out.

Units of measure. Until the United States gets in step with the world, writing that includes the "English" system of measurement (which the English no longer use) will need to be translated into the metric system (also called the Système International, or SI). What's tricky is that one or the other set of measurements will be "off." There is no exact equivalency between units in the two systems, such as inches and centimeters, feet and meters. If it's crucial that your readers get the amounts right, you will need to measure in both systems, rather than using a conversion formula. If a formula will do, there's some inexpensive shareware that you can install on your computer to take care of the math. Don't think it matters? Quick: If you're driving 60 kph, are you speeding or creeping? If it's 20°C, are you hot or cold?

"We." We, the people of the United States, have a tendency to speak and write as if we were everyone. The words "America" and "Americans" grate on readers in Canada, Mexico, Central America, and South America when they're applied only to the United States and its citizens. Changing the phrase "this country" to "the United States" makes your writing clearer.

WRITING ABOUT THE WEB

Print publications tend to have a style in place. The problem that arises in writing about the Web is that it introduces elements not covered in the existing style. Were there no style, there would (ostensibly) be no problem—at least, not until the second time you referred to a Web site, only to find that it had been done differently the first time. Once you find a second reference, you need to establish a style.

References to Web Sites

The term "visit" is often applied to Web sites: "To see the pandas, visit www.sandiegozoo.org." It isn't really necessary to say "visit the Web site," since the format of the URL makes it clear that you're not talking about booking a trip to San Diego. The preposition "at" is common: "A useful list of references can be found at www.libraryspot.com." You may also use "on" in the same context: "You can find just about any book on www.amazon.com."

Citations

More and more information is available in electronic format. Many sources are offered in both print and electronic form; some exist only as CD-ROMs or Web pages. The principle for citing electronic references remains the same as that for citing print: Provide complete information that will enable the reader to locate the source.

The amount of information you need to give in text references will depend on whether you provide a formal bibliography. If you do, your in-text citation should resemble that for a print reference: give the author (or title, if no author is available) and year of publication. For example, (Bunn, 2000) or ("A Beginner's Guide to HTML," 1996). Follow these guidelines for bibliographic entries:

Web sites. Attribute material to the author or authors (if you can identify them) and then give the title, URL, and the date you accessed the Web site. The latter information is important because Web sites change frequently.

> Bunn, Austin. "Prisoner of Love." http://salon.com/ent/feature/2000/01/27/letourneau/index.html (Jan. 27, 2000). [in this example, the URL includes the date the article was published online]

"The Living Pond." www.sover.net/~bland/pond.htm. (Jan. 27, 2000). [this example has no author listed]

Bray, Hiawatha. "DoubleClick's Double Cross." www.digitalmass.com/news/daily/0127/upgrade.html. (Jan. 27, 2000). [the URL includes the month and date, but not the year of publication]

E-mail. Cite e-mail messages as personal correspondence. Do not include the e-mail address of your correspondent. Give the person's name (rather than the e-mail alias) if you know it. Where you would list the title of a publication, list the subject line of the e-mail message. Give the date the message was sent.

Smith, John. "Woodchuck population in Rhode Island." June 30, 2000. Personal e-mail.

Jsmith. "Woodchuck population in Rhode Island." June 30, 2000. Personal e-mail. [in this case, you know the correspondent's alias but not the real name]

Electronic lists and newsgroups. Because these lists are available to the public, messages received through them are not considered personal correspondence. (They are, however, subject to copyright protection. See Copyright Issues on page 34.) In addition to the information you would cite for an e-mail message, include the name of the list or newsgroup, and both the date the message was posted and the date you accessed it.

Baughman, David. "Re: [Q] A-Fib and cardioversion?" Jan. 21, 2000. AT&T WorldNet Services, sci.med.cardiology. (Jan. 27, 2000).

CD-ROM. List the author's name, the title of the section or article, the title of the CD-ROM, and any other publication information that is available. Since CD-ROMs are less mutable than Web sites, do not list the date you accessed the material.

"A Beginner's Guide to HTML." *Microsoft Works & Bookshelf 1996-97.* Microsoft Corp., 1996. [no author]

You can deal with electronic references in electronic publications by providing a link to any references you cite. Such links eliminate the need to provide URLs, since readers can access the reference directly. "Jack Powers suggests that your Web site must attract search engines in order to attract readers" is the Web substitute for "Jack Powers, at www.electric-pages.com/articles/wftw2.htm (Feb. 2, 2000), suggests…." Bear in mind, however, that Web documents are transient. The link may not take readers to the article you have in mind indefinitely. Providing the URL is a more durable form of citation.

Copyright Issues
Material published on the Internet is intellectual property, and as such is protected by copyright. The fact that it's easy to copy the information does not alter the principles of ownership, any more than does the fact that photocopiers make it easy to copy printed materials. Even material posted on electronic mail groups is subject to copyright protection, although, since it may have no commercial value, the author of the material would have difficulty recovering financial damages for copyright violation.

When you publish a document on the Internet, it's a good idea to include a copyright notice, just as you would for a print publication. Doing so will make it easier to recover damages for copyright infringement. However, material that appears without such a notice is not in the public domain unless the owner specifically states that it is.

Remember, from a copyright point of view, publishing on the Internet is no different from publishing in print. The same rules of fair use apply. For a useful article, see Brad Templeton, "10 Big Myths about Copyright Explained," http://www.templetons.com (Feb. 2, 2000).

WRITING E-MAIL
The creators of ARPANET messaging, precursor to Internet e-mail, published a paper in 1978 in which they approved of how "one could write tersely and type imperfectly, even to an older person in a superior position and even to a person one did not know very well, and the recipient took no offense."

There's some debate today about how much typos and poor grammar matter in e-mail messages. Many of the Web's founders, themselves articulate standard-bearers for civilized communication, nevertheless accept that e-mail is created so quickly it is liable to formal errors. One editor recounts that an author with whom she was corresponding, the Jargon File's own

Eric Raymond, commented graciously, "I never take offense at mistypings in e-mail. I don't take offense because as long as I know what you mean I don't even really see the errors."

It's true that e-mail is a communication tool, not a literary genre. It has conventions, just as other forms of communication (letters, phone calls) do. How you write e-mail depends largely on context. If the only e-mails you send are to your children at college, you don't need to read this section. But if you send business e-mails as well, take another look at what the recipient may be expecting.

Use the Subject Field
Most e-mail systems display the sender's name and the subject field before the reader actually opens the message. Try to make your subject field as informative as possible, so a reader faced with a dozen e-mails first thing in the morning will have some idea which to open first.

Keep It Brief
People are receiving more e-mails every day. Opening and reading one message may take less than a minute, but the minutes add up fast. There's no need to be telegraphic—you're not being charged by the word, after all—but do respect your reader's time. E-mails are like the interoffice memos of yore. Each should address a single topic, and do so succinctly and clearly.

Don't Be Cute
Continuing with the interoffice memo analogy, don't bother with a salutation or a complimentary close. The program will tell the recipient who sent the e-mail. There's a tendency to use sig (signature) files, which automatically add a short message to the end of an e-mail, like those clever answering-machine messages people found so entertaining a few years ago. The first time you called John Doe and heard his answering machine imitate Richard Nixon, you might have been amused. The third time you called, you were probably ready to yank the phone out of the wall and throw it at him. On the other hand, it may be useful to include your phone number and postal address in a sig file, in case your correspondents need to reach you in one of these old-fashioned ways.

And another thing…skip the emoticons. Those are the combinations of characters that let you make something that, held sideways, looks vaguely like a smiley face, a wink, etc. Some folks claim that readers won't know they're being facetious unless they add an emoticon. Nonsense! For centuries, writers have been able to convey humor perfectly well using words alone. If you doubt your ability to do so, just dispense with the humor (not a bad idea in business communications of any sort).

Wait a Second
Once you hit "send," that e-mail is gone beyond recall. Before you send it, take a moment to read through it. Is everything spelled correctly? Have you covered your subject clearly? Most important, have you said anything you'll regret? And check the address box! It's easy to click on an address one line above or below the one you're aiming for, and the consequences can range from embarrassing to horrifying. Or you can accidentally reply to an electronic mailing list posting by a friend of yours and end up posting to the entire list—perhaps slamming other list members in public. The ouch really stings.

WORKING WITH WORD

Microsoft Word is the word processing program in widest use today. As its developers update it, they keep adding features that are meant to make users' lives easier. It's a matter of opinion whether they do or not. We were quick to disable the leering paper clip icon that popped up when we typed the word "Dear," saying, "It sounds like you're writing a letter. Would you like some help?" (Yes we are, and no we wouldn't.) On the other hand, we're bad enough typists that we don't mind having spelling errors pointed out as we go.

Word has a setting that enables it to "recognize" URLs or e-mail addresses and create links from a Word document to the Internet. That can be handy, but it has confused some users, especially those who don't want to print URLs with an underscore. You can turn this setting (along with several others) on or off by clicking Tools, then choosing AutoCorrect. The option is under "Autoformat as you type." If the "Internet and network paths with hyperlinks" box is checked, click that box to uncheck it.

The key thing to remember is that whatever Word does automatically is an option that you can turn on or off. The trick is to find out where those options are located—they aren't particularly intuitive. You can find them through trial and error; ask a coworker; click Help; or refer to printed documentation. We've listed the methods in order of preference, but we've used them all.

2. Quick Reference List of Troublesome Terms

As new technology seeps into our homes, cars, offices, and everywhere we play and travel, we can find ourselves suddenly trying to use new media and IT terms "properly" in context without any idea what rules apply. How can you tell the difference between true jargon and acceptable idiomatic expressions, when they're not written in the idiom you're personally familiar with? If you see the phrase *home edutainment center,* is it jargon to be edited (*home educational entertainment center* is pretty clunky) or will people understand what it means?

Along with every other industry, print publishing—traditionally somewhat conservative and not terribly fond of technology for its own sake—has been dog-paddling until recently in the riptide of technological change. And naturally enough (but regrettably), many Internet publishers still have more interest in how well graphics, links, and information architecture work than in whether copy is hyphen-bald. Somewhere in between these conservative and radical zones, most of us are trying to strike a precarious balance.

RULES, ADVICE, YOUR BEST GUESS
In this troubleshooting glossary, EEI Press editors offer advice on several hundred terms that are often encountered and misunderstood. We offer firm recommendations whenever our review of printed and online publications

and recent updates of major style references makes consensus possible. When there are alternative forms, we show you the acceptable options. When major authorities disagree, we note who advises what—at least, right now—so you can make decisions that will be relatively consonant with your main style guide. (See the bibliography, page 111, for examples.) Sometimes all we can do is help you make your own best guess about what will suit your audience.

Many of these words are up for grabs, and probably within months—or tomorrow—their form is likely to be restyled and their use somehow altered. Variant forms are on their way to convergence sooner or later, one way or another. The one thing you can count on is the steady evolution toward *downstyle*, a trend toward omitting optional but unnecessary capitalization and punctuation. The term is an example of itself.

We have to keep watching new media terms. They're hyperkinetic. Now that we've written the definition of *repurpose*, we'll probably decide eventually that it should have been flagged as tedious jargon for *reuse* or *recycle*. As the new gets old, some older, simpler terms may start looking better and better to us (now that we understand what we're saying—a serious factor).

That's why even as we've been researching this guide, we've seen *e-mail* trying to become *email*. Why add a hyphen for its own sake? Well, of course, there *are* reasons to add it. (For the full discussion of *e-* words and *e-mail*, see pages 16 and 34.) The point of style in general is to impose consistency. The point of downstyle in particular is to simplify what is already correct on the premise that even simpler is even better—even more readily accessible to readers.

Up with downstyle….

WHY ARE THESE PARTICULAR TERMS HERE?

This is a consensus guide; the glossary terms are widely used, in enough mainstream publications, so that a majority of writers and editors are likely to encounter them when writing about new media. These are also terms that people working in new media who write for the public need to use consistently—so readers don't become too distracted or annoyed to take in new (and sometimes conceptually difficult) information.

Plenty of excellent, comprehensive, microscopic computer dictionaries are available; we're not trying to create another one. In fact, this isn't strictly speaking a glossary, which is "a list of difficult or specialized words with their definitions." Not all the terms are defined; in many instances, it wouldn't help you much if they were. We presume you have other resources for verifying the accuracy of technical subject matter.

Instead, use this section as a quick check on a range of the most common challenges pertaining to IT and new media terminology. These are words to watch. That's why periodic updates of this glossary will be available at the EEI Press Web site. (For more information, visit www.eeicommunications.com/press.)

These are the main categories into which the selections in this glossary fall:

- Terms with idiosyncratic spelling for which all mnemonics fail. You just have to look up *eBay* so you can get it right for an article you're writing about online auction sites.

- Preferred, acceptable, evolving, and problematic usage of new terms as nouns, verbs, and modifiers. You don't really want to be one of those people with pursed lips rapping the knuckles of your coworkers for saying "Send me an e-mail" instead of "Send me an e-mail note," do you? Or will you cheer them on to all sorts of *e*-verbs such as *e-leverage*? (Yeah, noun, too.)

- Terms with arbitrary style treatments, including commonly used abbreviations, acronyms, and initialisms. A good number of new media words, especially the notorious, glorious *e-, i-,* and *cyber-* words, are coined for marketing purposes or are proprietary names and trademarks.

- Terms that are best left unchanged—even if they strike you as too informal or jargonish—as long as they are clear in context. It marks you as old-fashioned to turn every Web- and computing-related idiom into Formal Plain English. We found that words we considered on the fringe as we began this guide had already become household words a few months later—*e-tailing,* for example.

When corporate publications, marketing materials, and Web site content contain a lot of special terms and reflect the style preferences of a lot of different authors, it's important to create and enforce an in-house style guide.

Key areas in which to make consistent decisions include capitalization, punctuation, spelling, and preferred usage for enforcing trademarks and other proprietary names. For example: "Don't use *e-mousetraps* as generic shorthand; use *eMousetrap rodent control devices*."

(By the way, when we wrote that mousetrap example, it was imaginary. But we did a search and discovered that it's indeed a trademark. Scary how thin the dividing line has become between the real and the theoretical.)

One definition of *style* is "The fashion of the moment." Another is "A slender, pointed writing instrument used by the ancients on wax tablets." Another is "a customary manner of presenting printed material, including usage, punctuation, spelling, typography, and arrangement." Yet another is "the usually slender part of a pistil, situated between the ovary and the stigma." But wait: "a surgical probing instrument," too! Little words mean a lot; a little carefully applied style goes a long way toward fruitful, clean prose that does its work.

That's all style is for. Not to worry editors (or give them a truncheon to wield, either). It's really for readers. And it works by not getting in the way.

GLOSSARY

A Note on the Entry Format

Terms boldfaced in text are cross-references to related entries elsewhere in this glossary. Sample sentences showing use in context are enclosed in quotation marks. Italics are used for additional examples of the entry term and for the spelled-out terms that abbreviations and acronyms stand for.

*nix *Jargon.* (adj.) **nix* is used to mean some or any form of Unix. For example, "They run VM, MVS, NT, and lots of **nix* servers."

24/7 *Jargon.* (adj.) Used to indicate a service or source that's available around the clock, 24 hours a day, seven days a week. "Our customer service department is open *24/7* to serve you." Also written *24 x 7, 24 by 7,* and *twenty-four seven* (the "Tina Turner *Twenty-Four Seven* Millennium Tour").

access (n.) To have or gain access refers to the privilege or right to use a computer resource. "All employees have *access* to the company intranet." (v.) *To access* means to be able to gain access to data, a system, or processes. "You can *access* that file on the Y: drive." (adj.) Used as an adjective, *access* appears as a separate word, not hyphenated or compounded: *access code, access control, access line, access mechanism, access method, access path, access privileges, access provider, access right, access server, access time.*

accessible (adj.) *Accessible* features and operating controls have been designed for easy use by people with disabilities (visual, physical, or mental impairments). In a new-media context, *accessible* may also describe the degree to which features can be taken advantage of by nonnative English speakers and Internet users from developing countries who have very slow dial-up speeds. (n.) *Accessibility.*

add-on / add-in (n.) An *add-on* is a peripheral attached or added on to a computer, such as a modem or a printer. An *add-in* is a feature that enhances an existing piece of software. (adj.) Both *add-on* and *add-in* can be used as adjectives with an appropriate noun. For example, an *add-on board* is a circuit board that is affixed to an expansion slot on a computer. See also **plug-in.**

adopter, adapter (n.) An *adopter* is someone who embraces or takes up the use of a technology, idea, or practice at some point in its development, either early or late. An *adapter* is a circuit board or device that makes it possible for a computer to use a peripheral device such as a scanner. Technology and e-commerce reports often discuss the role of the *early adopters*—people who embrace new technologies—as opposed to the mainstream users of a product, technology, or service. Also *adoption, early adoption.* (v.) *To adopt.*

analog (adj.) New-technology writing often uses the contrasting terms *analog* and **digital.** Most of creation breaks down into either *analog* or *digital,* and as the world sweeps toward an increasingly digital age, it is important to know the difference. *Analog* processes represent data as variables that vary continuously rather than discretely. *Digital* processes are represented by numerical values—at the most basic, 1 or 0 (for on or off). They also represent a finite number of values. For example, a *digital watch* display leaps from one number to the next; it displays a finite number of variables (times of the day). In contrast, an *analog watch* (the old-fashioned kind with hands that move) represents an infinite number of variables as its hands sweep around the dial. Most modern computers are *digital;* human beings are *analog.*

analog-to-digital converter (ADC) (n.) Hyphenated. Also called an *A-D* or *A-to-D converter*.

anti-aliasing *Anti-aliasing* is spelled as two hyphenated words to avoid the awkward double vowel construction. (n.) *Anti-aliasing* is a process that removes jagged edges ("aliasing") from onscreen type and images containing lines or curves. (v.) As a verb, *anti-alias* is also spelled as a hyphenated compound: "Graphics that you have *anti-aliased* look better onscreen." The word is also (less commonly) spelled closed up as *antialiasing* for both noun and verb.

application (n.) In technical writing, an *application* is an executable program that performs a specialized function such as word processing or spreadsheet calculations. In most contexts, it is preferable to use the familiar word **software**. An *application* is designed for end users. A contrasting term is **utility**, which refers to an executable program that does system maintenance. A *killer app* is an application that defines a category, revolutionizes an industry, or kills off the competition. "Lotus 1-2-3 was the killer app that launched the popularity of personal computers for small business." A type of *application* is a **client**.

ARPANET (n.) The first wide-area computer network developed by the U.S. Department of Defense in the 1960s that became the precursor of today's Internet. *ARPANET* means *Advanced Research Projects Agency Network*. Also spelled *Arpanet*.

ASP (n.) 1. An *application service provider*. An *ASP* is a business or nonprofit organization that rents software applications across a wide-area network. It is a way for companies to outsource the management and maintenance of software applications. 2. *Active Server Pages*. *ASP* is one of several proprietary technologies that enable generation of dynamically created Web pages. See also **dynamic**.

authoring *Jargon.* (v.) To *author* still means to write a book. However, *authoring* has the new-media connotation of creating an online application such as a **multimedia** presentation, help system, or computer-based training program. (adj.) Common *authoring tools* include Director, Authorware Attain, and Instructor. Also *authoring system, authoring language, authoring program*.

b-to-b *Jargon.* (adj.) *B-to-b* is short for *business-to-business,* meaning business that is transacted between businesses, rather than from business to consumer. This abbreviation is also spelled *B-to-B* and *B2B.*

b-to-c *Jargon.* (adj.) *business-to-consumer,* also spelled *B-to-C, B2C.*

back end (n.) Use two words. This is a server or program that processes data received from a **front-end** application. It is common for online forms to have a database as the *back end.* (adj.) Hyphenate the adjective, as in *back-end processor* and *back-end infrastructure.*

backup (n.) A duplicate copy of valuable data. (v.) To *back up.* The process of making a *backup.*

backward slash or **back slash** (\ ) (n.) See **slash.** Use two words rather than *backslash.*

bandwidth (n.) *Bandwidth* is a measure of data transmission capacity. Higher *bandwidth* allows for faster transmissions and also for transmission of a greater volume of data. *Bandwidth* in a **digital** system is measured in bits or bytes per second. In an **analog** system, *bandwidth* represents the range between the highest and lowest frequencies transmitted and is measured in **hertz** (formerly called cycles per second), abbreviated as *Hz.* In business jargon, *bandwidth* measures the maximum capacity of a business or factory to produce results: "Apple Computers has foundered in the past because it did not have the *bandwidth* to keep up with customer demand for its products."

bandwidth on demand (adj.) Use no hyphens.

BBS bulletin board system

Bezier (adj.) A computer graphic term. *Bezier curves* are based on formulas developed by a Frenchman, Pierre Bézier. In common usage *Bezier* is spelled with an initial cap, but without the accent mark.

bitmap (n.) A *bitmap* is a pattern of dots. Together, the dots make up an image. The image itself is also called a *bitmap.* (adj.) *Bitmapped* graphic formats include bmp, jpeg, gif, pcx, and tiff. Also *bitmapped image, bitmapped graphic, bitmapped font.* Alternative though less common spellings are *bit map* for the noun and *bit-mapped* for the adjective. *Bitmapped graphics* are also called raster images or raster graphics. A painting program, which stores images as pixels, is used to create and edit *bitmapped* images. See also **vector graphic**.

bot (n.) A *bot,* short for robot, is a program that automates repetitive tasks that would otherwise be done by a human. For example, a *chatterbot* interacts with customers on an e-commerce site, eventually completing the sale. Another type of *bot* continually accesses Web sites to create an index that will be used by a search engine. *Bots* get around; other types are the *knowbot, sales bot, shopping bot, spambot,* and **spider**. Also known as an *online personal assistant, intelligent agent,* or *agent.*

bricks-and-mortar (adj.) In cyber-jargon, *bricks-and-mortar* (seen almost as often as *brick-and-mortar*) usually refers to a physical office, building, or retail store—as opposed to an operation existing exclusively on the Web. See also **clicks-and-mortar.**

broadband (adj.) It's one word.

browser (n.) A *browser* is a software application used to scroll through and view information contained in documents or a database. A *Web browser,* such as Netscape Navigator or Microsoft Internet Explorer, is used to view HTML-formatted documents on the World Wide Web. *Browser* **plug-ins** are required to view some multimedia file formats that add sound and video to a Web page. Usage note: Users look at Web pages *with,* not *through,* a Web *browser.*

Bubble Jet (adj.) A *Bubble Jet printer* is an **inkjet** printer made by Canon. *Bubble Jet* is a trademark of Canon, Inc.

burn (v.) The process of writing data to a CD-ROM is often called *burning a CD,* presumably as a nod to the laser technology that writes data to the disc by burning microscopic holes in it. (n.) *Jargon.* A *CD burner* is a reference to a device, such as a CD-R device, that writes CD-ROMs. (adj.) A *burn-in period* or *burn-in test* is the stage during which computer equipment is run to ensure that all components work properly before it is released to market.

buyer pool, buyer pooling (n.) An online *buyer pool,* found on an e-commerce Web site, appeals to individuals who want to pool orders with an anonymous group of people to obtain volume buying discounts. As more individuals join the *pool* for a specific item, the price goes down.

A *pool* is usually open for a fixed period. At the end of this time, an order is placed and the number of buyers will determine the final selling price. Also called *aggregated buyer pooling, online buyer pooling,* and *reverse auction.*

CAI *computer-assisted instruction* or *computer-aided instruction.* See also **CBT**.

CBT *computer-based training.* (n., adj.) *CBT* programs are interactive, computer-based instructional programs. *CBT* is also called *computer-assisted instruction* (**CAI**).

CD-I *CD-Interactive.* (n., adj.) A type of compact disc developed by Sony and Phillips that holds video, audio, data, and graphics. The *CD-I* does not play on a CD-ROM device. This format has not achieved wide acceptance.

CD-R *CD-Recordable.* (n.) A *CD-R* is the compact disc used in a *CD-R drive.* This disk can be written to only once (adj.) A *CD-R drive* is a type of CD-ROM drive that can read and write CDs. The *CD-R drive* can record each disc only once.

CD-ROM *compact disc read-only memory.* (n.) *CD-ROM* used as a noun refers to the disc itself. (adj.) As the acronym includes the word *disc,* it is repetitive to say *CD-ROM disc.* Correct use as an adjective: *CD-ROM drive* or *CD-ROM device.*

CD-RW *CD-Rewritable.* (n.) A *CD-RW* is the compact disc used in a *CD-RW drive;* it can be erased and rewritten many times. (adj.) A *CD-RW drive* can read, write, erase, and rewrite a *CD-RW disc.*

cellular phone (n.) This is a telephone for use in a cellular communications system. Synonyms include *cell phone, cellular telephone, digital phone, handheld phone, hand phone, mobile phone, walkabout phone,* and *cell. Cell phone* is commonly accepted; *cellular telephone* is a bit formal.

chat words (adj.) *Chat* words (such as *chat room, chat mode,* and *chat window*) show few signs of becoming closed compounds. In a few cases *chat room* is spelled as a closed compound, but there are far more occurrences of the word as an open compound.

check box (n.) Use two words. In an online form, a *check box* is a small box that displays a check when the option associated with it is selected. See also **radio button.**

checklist (n.) It's now a closed compound, as we bet *checkbox* will be someday, too.

CIO *chief information officer.* (n.) That's the top corporate executive officer in charge of information systems (computers).

click (n.) One depression of a **mouse** or pointing device button. (v.) To select an onscreen option by depressing a mouse or pointing device button. See also **double-click.**

click and drag (v.) *Click and drag* is a common expression in documentation to describe what a user does with the **mouse** on a computer: "*Click* on an item and drag it to a different part of the screen." Use no hyphens.

clicks-and-mortar (adj.) *Clicks-and-mortar* is a play on **bricks-and-mortar**. A *bricks-and-mortar* store has a tangible physical presence, as opposed to a store operating totally online. *Clicks-and-mortar* has most often been used to refer to businesses that have both an online and a physical (storefront) business presence. But it's also used to describe retail stores that feature Web kiosks and make Web site transactions available to customers while they're physically in the store. "*Clicks-and-mortar* companies offer consumers several options for merchandise returns."

clickstream (n.) The pattern of mouse clicks that a user makes while working on a computer or navigating through sites on the Web. The *clickstream* shows, for example, the path that a user took to navigate through a Web site or between Web sites.

clickthrough (n.) *Clickthrough* is a Web advertising term for when a person clicks on an an Internet banner ad and triggers the associated hyperlink that leads to more information. (adj.) The *clickthrough rate* refers to the percentage of viewers of a Web banner ad who clicked on the ad. See **Web traffic and advertising terms.**

client (n.) A *client* is an **application** that runs on a workstation or desktop computer but accesses a server for some functions. (adj.) *Client application, client software.* "E-mail software is the most widely used client application."

coaxial Use it as one word. (adj.) *Coaxial cable,* a type of wire used for cable television networks and computer networks. (*jargon,* n.) *Coax* is sometimes used as a shortened form of coaxial cable. Less common spellings are *co-axial* and *co-ax.*

co-locate (v.) In the wired world, to *co-locate* one's Web server means to locate it physically on another company's Internet-connected network. (n.) *co-location.* (adj.) *a co-location facility, co-location service.* Sometimes spelled *colocate* or *colocation.* This word is frequently misspelled as *collocation,* which means instead to place or group together.

CompuServe (n., adj.) The *s* is capitalized.

computer-to-plate (CTP) (adj., adv.) Note hyphens. See **direct-to-plate.**

convergence (n.) In the realm of information technology, *convergence* refers to the combining of computer, communications, and consumer electronics technology. (adj.) Examples of *convergent technology* include interactive television and cellular telephones with information-sharing features.

CPM *cost per thousand.* See **Web traffic and advertising terms**.

cracker (n.) One who develops malicious computer programs such as viruses and Trojan horses, pirates software, or breaks through computer security systems to access others' computers with malicious intent. Also known as a *black hat hacker* (as opposed to a *white hat hacker* who purports to use hacking talents only for nonmalicious activities). The media often uses the well-known term **hacker** when reporting on what are really the activities of *crackers.* (v.) *To hack* or *hack into.*

CRM *customer relationship management*

cross-media (adj.) *Cross-media* products simultaneously involve multiple media such as print, Web, and other publishing formats.

cross-platform (adj.) This describes a software program that works equally well on various computer platforms such as PC, Mac, and Unix.

cross-post (v.) To *cross-post* means to **post** the same message to several different newsgroups or mailing lists at about the same time.

CTO *chief technology officer* or c*hief technical officer.* (n.) This is the top corporate executive officer in charge of technology.

cyber- words Originally from *cybernetics,* the cut form *cyber* is usually attached to words in order to add the cachet of being computerized, electronic, or Internet-based. Most *cyber-* terms are closed compounds such as *cyberbuck, cybercafe, cybercash, cybercommerce, cyberculture, cybercop, cyberfeminism, cyberlawyer, cybernaut, cyberpiracy, cybersex, cyberspace, cybersquat,* *cybersurfer, cyberworld*. At times, open compounds are clearer and more attractive: *cyber economy, cyber investor*.

data (n.) In purely statistical or scientific usage, *data* is a plural (or count) noun meaning "pieces of information." *Datum* is the singular. But *data* is widely and increasingly acceptable in a collective (or mass noun rather than count noun) sense used with a singular verb. That's using *data* to mean "cumulative bits of information." When *data* means "information stored electronically," we recommend using *data* in its singular sense, with a singular verb: "The *data was* restored from the backup disk." Certainly we would say, "The *data is* corrupt" rather than "*These data* are corrupt," unless we are statisticians.

data- Compounds formed from the word *data* are almost always open whether used as nouns, verbs, or adjectives: *data bank, data bit, data bus, data center, data-driven processing, data entry, data field, data file, data flow, data link, data mart, data mining, data offset, data packet, data point, data rate, data set, data sink, data stream, data warehouse*. Notable exceptions are **database,** database administrator, database engine, datacom, datagram.

database It's a closed compound. (n.) *Database* can refer generally to a collection of electronically stored data that is available on demand ("Our company's most valuable asset is its massive online knowledge *database*") or to a set of files that are created and managed by a specific technology—a *database management system* (DBMS). (adj.) *Database system*.

DeskJet (n., adj.) *DeskJet* is the name for a line of **inkjet** printers made by Hewlett-Packard. The name *DeskJet* is a registered trademark of HP.

desktop (n.) 1. In a **GUI,** the *desktop* is the computer's onscreen work area where icons and menus allow the user to access all programs and functions. Also called an *electronic desktop*. 2. *Jargon*. A *desktop* is also a computer that is small enough to fit on top of a desk or in an individual's work area, short for *desktop computer* or *desktop model computer*. (adj.) 1. Pertaining to the GUI desktop: *desktop level*. For example, "All of the programs and functions on a computer can be accessed from the desktop level." 2. Activities and technology designed to be accessible from the user's business desk: *desktop computer, desktop conferencing, desktop publishing, desktop video*.

desktop publishing See **DTP.**

DHTML *Dynamic HTML*

dial-up (adj.) It's two hyphenated words. Examples: *dial-up connection, dial-up access, dial-up line, dial-up network.* (v.) To *dial up.* To use a modem operating across an ordinary telephone line to establish a temporary connection between a device and another computer or a network. (n.) *Jargon.* Although *dialup* and *dial-up* are both used as nouns in tech jargon, careful editors are advised to recast *dial-up* as an adjective: a *dial-up service.*

digerati (n.) A play on the Italian word *literati, digerati* refers to the savants/cultural elite of the digital revolution. While the word also appears spelled *digirati* and *digiterati,* we prefer the phonetic and much more commonly used *digerati.*

digi- words *Jargon. Digi-* words are compounds derived from **digital.** They are generally spelled as closed compounds. For example: *digibabble, digispeak, digitocracy, digerati.* Note: *DigiCash* is a corporation whose product is the **eCash** electronic payment mechanism.

digital (adj.) As an adjective, *digital* adds the loose connotation that something is used on a computer or is computerized. Examples are *digital cash, digital certificate, digital money, digital network, digital photography, digital recording, digital signature, digital video.* For a comparison of digital and analog processes, see **analog.**

digital money (n.) An anonymous, secure electronic payment system. Systems for electronic payment transfer include *digital coins, digital checks, digital wallets,* and *digital cash.*

digital paper See **electronic paper.**

digital stamp (n.) A *digital stamp* is postage that can be paid for online, then downloaded and printed onto labels or envelopes from a home computer. (adj.) Several companies offer *digital-stamp services* under different names, such as **E-Stamp**'s Internet Postage and Neopost's PC Stamp.

digital subscriber line See **DSL.**

digital-to-analog converter (DAC) Use it hyphenated. Also called a *D-A* or *D-to-A converter.*

digitize (v.) To convert an image, text, or a signal into **digital** code using a scanner or converter. "The 16 mm film of Kennedy's assassination was *digitized* by the National Archives."

directory (n.) In the context of the Web, a *directory* is a hierarchically organized index to sites and content on the Web. In contrast to most search engines, which are computer-generated, a *directory* is created by people who select and organize Web offerings by categories. Yahoo! began life as a *Web directory*, offering the first major hierarchical index of the Web. AltaVista began as a Web search site based on a powerful **search engine**. Today the distinction between a Web search site and a *Web directory* is less clear, because many sites offer both features—for example, Yahoo! and AltaVista.

direct-to-plate (adj., adv.) *Direct-to-plate* is used to indicate a pre-press technique in which digital files are burned directly to plates for printing, bypassing the step of creating a film negative. Less often seen as *direct to plate*.

disc, disk (n.) *Disc* is an alternative spelling for *disk*. The storage medium determines the correct spelling to use. Most authorities agree that most *magnetic computer disks* are spelled with a k (*floppy disk, hard disk, magneto-optical disk, RAM disk, diskette*) and most *optical disks* are spelled with a c (*video disc, compact disc, laser disc, digital versatile disc*). A number of the sources we reviewed did not distinguish between *disc* and *disk*, however, using them interchangeably. From our review, it is a bit premature to blur the lines that far. Perhaps the easiest generalization that will serve most writers and editors is that CDs are *discs* and all other types of storage media are *disks*. (adj.) *disk drive, disk farm, disk mirroring, disk operating system* (DOS), *disk partition*. See also **CD-ROM**.

disintermediation (n.) This means "the elimination of the middleman." One of the theories about how Web commerce will change the world is that it will lead to the *disintermediation* of retail sales.

DNS An acronym for both *domain name system* and *domain name service*.

domain name (n.) This is an Internet address, such as eeicommunications.com. A *domain name* corresponds to a numerical **IP address** that serves as a routing address on the Internet.

dot-com *Jargon.* (adj.) A *dot-com business* derives most if not all of its revenues from Internet-based sales and services. These businesses generally have corporate names tied to Internet addresses that end in *.com,* the extension that indicates a company. The word isn't in dictionaries yet, and it's a great example of a new term that lends itself to creative spelling. *Dot-com* is also spelled *dotcom* and *dot.com,* and, literally, the slightly passive-aggressive *.com* (it's easy to miss that little "dot" on a quick read or forget to pronounce it "dot"). Among the major business and news publications we surveyed, *dot-com* was the preferred spelling, and we prefer it, too. (n.) *Dot-com* also appears as a noun, as in: "The proliferation of profitable IPOs by *dot-coms* has Wall Street sharks in a feeding frenzy." (v.) We've also observed *dot-com* in the wild as shorthand for the act of formally registering a **domain name**: "When are you going to *dot-com* your company?" But that's too ambiguous; it could also be taken to mean "When are you going to take your **bricks-and-mortar** business online?"

dot-org *Jargon.* (n., adj.) This is a company, nonprofit organization, or institution possessing a Web address ending in .org. This organization may or may not have an offline (**bricks-and-mortar**) presence. (See the sidebar on page 22.)

double-click (v.) A hyphenated compound; *double-clicking, double-clicked. To double-click* is to depress the **mouse** button twice in rapid succession. The verb *double-click* is not followed by *on* so use "*Double-click* the AOL icon," rather than "*Double-click on* the AOL icon." (n.) Sometimes seen as an open compound: "*In Microsoft Windows and the Macintosh interface, you can use a double click to open files and applications*" (PC Webopædia). But we recommend hyphenating it.

download (v.) It means to receive data from a remote computer over a network or Internet connection. (n.) *Download* is also used as a noun: "Our Web site features a free *download* for people who register." This use verges on jargon, although it has become increasingly common. For now, we recommend recasting the sentence; for example: "People who register on our Web site can *download* a free electronic publication." Be aware of the difference between *download* and **upload.** A file is *downloaded* when it is received from a remote location. You *upload* a file when you transmit it to a remote location.

drag and drop *Jargon.* We've seen this common computer expression used as a verb, adjective, and noun. (v.) While the expression "*drag and drop* the icon into the trashcan" hurts the ears, *drag and drop* clearly expresses an instruction for a specific action. In the *Microsoft Press Computer Dictionary* we find the following more palatable—if lengthier—example: "To delete a document in the Mac OS, a user can *drag* the document icon across the screen *and drop* it on the trashcan icon." A similar expression is **click and drag.** (adj.) When *drag and drop* is used as an adjectival compound, we hyphenate it, as in the sentence "Both the Windows and the Mac OS offer *drag-and-drop functionality.*" (n.) *Drag and drop* is also used as a noun. In the *Free Online Dictionary of Computing* we found the following example: "The biggest problem with *drag and drop* is does it mean 'copy' or 'move'?" See also **mouse.**

DSL *digital subscriber line.* (n.) Types of DSL include *ADSL* (asymmetrical DSL), *SDSL* (symmetric DSL), *HDSL* (high-data-rate DSL or high-bit-rate DSL), *RADSL* (rate adaptive DSL), and *VDSL* (very-high-data-rate DSL).

DSLAM *digital subscriber line access multiplexer*

DTP *desktop publishing.* (n., adj.) The line between DTP and word processing has become moot as word processors acquire more sophisticated page layout features. The distinction we usually draw is that the purpose of *DTP* is to prepare professional publications for commercial printing. Word processors are well-suited to preparing documents for office laser printer output. At the low end, however, home *DTP packages* allow users to create nicely embellished publications like greeting cards and certificates for output to an inkjet printer. At the high end, word processors can be used to produce books for commercial reproduction and binding on a DocuTech machine. (It may no longer be necessary to spell out *DTP* on first use, but it's a courtesy.) It's also advisable to begin a sentence with the spelled-out form (as for many other acronyms).

DTV *digital television.*

DVD *digital versatile disc.* (n.) Originally known as the *digital video disc,* this disc looks like a **CD-ROM** but holds much more data, therefore allowing full-length movies and audio programs to be recorded on the disc. *DVD* refers to the disc itself. Also common is *DVD-ROM*—this is a read-only *DVD.* A *DVD* can be recorded on both sides and holds several gigabytes of data per side; new technology will increase the storage capacity to over eight gigabytes per side.

The device that plays the *DVD* is called a *DVD drive* (or *DVD-ROM drive*) or *DVD device*. The newer *DVD devices* can play and record to both *DVDs* and CDs. (adj.) *DVD technology, DVD drive, DVD device, DVD player.*

DVD-R *digital versatile disc-recordable.* (n.) A write-once DVD.

DVD-RAM, DVD+RW (n.) Competing (and incompatible) standards for a high-capacity, rewritable DVD.

DVD-ROM *digital versatile disc-ROM.* (n.) A read-only DVD.

Dvorak keyboard A keyboard developed for greater speed and ease of typing compared with the performance of the traditional **qwerty** keyboard. The keys most often accessed by English-language typists are placed in the home row of the *Dvorak keyboard.*

dynamic (adj., adv.) *Dynamic actions* take place when they are needed rather than in advance. *Dynamic Web pages* are composed of components collected and served on demand or **on the fly. ASP** (Active Server Pages) is one of several proprietary technologies that enable generation of *dynamically created pages.*

e- words The prefix *e-* is short for *electronic. E- words* are catchy, popular, and probably the most abused category of words in the English language today. Words like *e-commerce, e-business,* and *e-mail* have sprung from nowhere into the mainstream in a matter of years. *E-What?* is titled to reflect the sometimes gratuitous incorporation of *e-* into general writing to mean "anything remotely technological."

Marketing copywriters, headline writers, and start-up companies have turned just about every conceivable word into an *e-word* to give it the cachet of "modern, new, and electronic." Readers may regard new *e-words* as trite rather than clever; exercise restraint in coining them.

For consistency, we prefer to spell *e-words* with a hyphen, and that includes *e-mail*. Recommendations for capitalization of *e-words* are found in Capitalization, page 7.

e-book, eBook (n.) Strictly speaking, an *e-book* is an electronic book. Be aware when you use the word that the electronic publishing industry is swiftly redefining what exactly an *e-book* means. The Open Electronic Book Forum has defined a common technical standard for the electronic book called the Open eBook Publication Structure Specification. Read more about this initiative at www.openebook.org. (adj.) To read an *e-book* usually requires an *e-book device* or *e-book reader*. A dedicated *e-book reader* is a **handheld** device designed for optimal *e-book viewing*. Note that *Rocket eBook* is a trademark of NuvoMedia.

eCash *eCash* is a registered trademark for the digital money system by DigiCash. If you're trying to use *e-cash* generically, are you infringing? Maybe.

editor (n.) A program used to create and edit text and code: *a text editor, line editor, HTML editor, source code editor*. An *editor* is less powerful than a word (or text) processor, which in turn is less powerful than a desktop publishing (**DTP**) program. A word processor usually has much more sophisticated formatting features than an *editor*. Also a *video editor*—a machine used to edit video. (v.) *To edit, editing* (code, HTML, etc.). (n.) A person who *edits*. (v.) To prepare for publication or public presentation. To alter, adapt, or refine especially to bring about conformity to a standard or to suit a particular purpose—"carefully *edited* the speech" (from *Webster's 10th*). *Editors* are not mere rule appliers. In the electronic age, they are called on more and more to be arbiters of the language where style guides and clearly defined rules do not exist.

electronic paper (n.) *Electronic paper* is an electronic display made from a thin sheet of flexible plastic. It is printed with electronic ink containing microscopic particles that change instantly in response to electrical charges and the same sheet can be reprinted thousands of times. Also known as *digital paper*. *ePaper* is a registered trademark of Adobe. Is *e-paper* an infringement? Maybe.

e-mail (n.) Short for electronic mail, *e-mail* used as a noun refers both to an electronic text message *(an e-mail)* and to the system of communication by electronic text messages over a network *(communicating by e-mail)*. (adj.) *E-mail message, e-mail system, e-mail traffic* (v.) To send messages by *e-mail*. Less common spellings are *email* and *E-mail*. Use *E-mail* at the beginning of sentences. Use two initial caps for title case (E-Mail). We've drawn a line in the sand by specifying *e-mail* rather than *email*. The trend, especially among computer industry publications, is toward the closed form. However, because *e-mail* falls within the group we've classified as *e-words*, we prefer to treat it consistently with words of parallel construction. *Ecommerce* and *ebusiness* are not as widely used as *e-commerce* and *e-business*, so we recommend keeping the hyphen in *e-mail*—for now. See also **rich e-mail.**

e-money E-Money, Inc., holds the tradenames EMONEY, E-MONEY, EMONEY.NET, E-MONEY.NET, and E-MONEY.COM, among other permutations on the term. Are you infringing if you're using it generically? Maybe.

EPS *encapsulated* **PostScript**

ERP *enterprise resource planning*

e-stamp (adj.) The word *e-stamp* is tempting to use as a generic term for electronic postage. A better, nonproprietary word is **digital stamp**. E-Stamp is a registered trademark and also a service mark of E-Stamp Corporation. Several other firms offer postage/digital stamps over the Internet, including Stamps.com and Neopost.

e-tail *electronic retail*. (n.) This word has been around for a few years; it popped into prominence late in 1999 as the U.S. financial and popular press closely monitored the performance of online merchants (e-tailers) during the Christmas season. (adj.) "The e-tail business."

Ethernet (n.) Ethernet is the most commonly used **LAN** protocol. Ethernet is always capitalized. It is a trademark of the Xerox Corporation. (adj.) "An Ethernet network."

eToys (n.) This is the sort of corporate name most likely to drive editors over the edge. The electronic age is bringing forth all kinds of corporate names with lowercased initial letters and midcaps. As we say on the back cover of this book, capitalize these names when they begin a sentence. The essential "new sentence" cue should not be superseded by cleverness.

E*TRADE E*TRADE is one of those names, along with Yahoo! and eToys, that makes us question how we want to style funky new corporate names—the way the company styles it, or the way that looks better/more acceptable to us in print. The company styles it E*TRADE. In practice, the name is usually written E*Trade by everyone outside the company. See Keeping Up with Style for a more detailed analysis of use of corporate trademarks.

extranet (n.) This is the portion of a corporate **intranet** that is available to customers/clients, vendors, suppliers, and/or business partners, usually employing password-protected access.

extreme *Jargon.* (adj.) A popular word, and definitely jargon, successor to *the bleeding edge*. Over the past several years, *extreme* has been used to mean "pushing the edge" or "taken to the ultimate extent" in many contexts—it's quickly becoming institutionalized as it's co-opted for marketing purposes. *Extreme sports* have lead to the *X-Games*. So *extreme* doesn't necessarily add the connotation of digital or electronic, although the Internet and all its associated technologies have pushed everybody to the *extreme edge* in one way or another (feeling *extreme* information anxiety anyone?). Anything that challenges the mind, imagination, or body is, idiomatically speaking, *extreme*. There will always be an *extreme edge*—maybe the word *extreme* won't be enough to describe it, though, as the hype inevitably becomes the norm, and then demoted to the equivalent of merely clichéd, *extremely cool*.

e-zine See **zine**.

FAQ *Jargon.* (n.) A *FAQ* is a list of *frequently asked questions,* usually posted on a newsgroup or Web site to provide a place to answer common questions asked by newcomers. Use of *FAQ* outside the Web and newsgroup context has become trendy but should be used carefully—not everyone will know what you mean. (adj.) A *FAQ list*. Pronounced "F-A-Q" or "fack."

fax Fax is short for *facsimile*. It does not need to be capitalized. Do not use FAX. (n.) Both the machine itself and the message received by fax transmission are *a fax*. (adj.) *A fax machine, a fax transmission* (v.) "I need *to fax* this memo immediately." As a transitive verb with an indirect object, it's slangy but inevitable: *Fax me your résumé.*

fiber optics (n.) *Fiber optics* is a communications technology based on data transmission on glass or plastic threads (fibers). (adj.) *Fiber-optic*. Hyphenate fiber-optic when it is used as an adjective: *fiber-optic cable, fiber-optic transmission. Fiber-optic cables have greater* **bandwidth** *than metal cables.*

file name (two words, n. and adj.)

File Transfer Protocol See **FTP**.

firewall One word as both noun and adjective. (n.) A computer network or a Web site can be protected from unauthorized users by a *firewall,* a protective barrier of hardware and/or software. (adj.) Such a network or Web site has *firewall protection.* (v.) *Firewall* can also be used as a verb: *This system has been firewalled.*

FireWire (adj.) *FireWire* is Apple's trademarked version of a high performance serial bus that conforms to the **IEEE 1394** Standard. This is an extremely high-speed **port** that can connect up to 63 peripheral devices to a computer. *FireWire* is often confused for a generic term, which it isn't. Other manufacturers make products that conform to the IEEE 1394 standard; the names just aren't as catchy or as memorable. Sony has its i.Link, Adaptec refers to its dry-as-toast "1394 products."

flame *Jargon.* (n.) A *flame* is a vicious comment or personal attack transmitted by e-mail or posted to a **BBS** or **newsgroup**. Also *flaming.* (v.) To send or post a flame; to be the recipient of flaming: *flamed for posting too many personal messages to the e-mail list.*

Flash (adj.) *Flash technology* allows designers to create files that deliver vector graphics, sound effects, and animation over the Web. The terms *flash* and *flash graphics* are often used as generic terms to mean any graphics-with-sound Web effect. In fact, *Flash* is a trademark of Macromedia and is properly used as an adjective, capitalized: *Flash technology, Flash player.* (n.) In common usage, however, *Flash* is used as a noun, even on Macromedia's Web site, as in the following quote from Macromedia's Flash FAQ: "*Flash* is the Web standard for vector graphics and animation. Web designers use *Flash* to create beautiful, resizable, and extremely compact navigation interfaces, technical illustrations, long-form animations, and other dazzling effects for their site. *Flash files* can play back with the **Shockwave** Player or Java" (italics ours).

font (n.) The difference between a *font* and a **typeface** can be confusing. In brief, a *font* is a set of characters for a particular typeface. A *typeface* is a design for a set of characters. Among the most common business-document *typefaces* are Courier, Times Roman, and Helvetica. Each *typeface* is made up of sets of *fonts,* which describe characteristics such as size, weight, and slant. Most *typefaces* include a minimum of four *fonts:* normal weight, bold, italic, and bold italic. For links to typography resources and examples of contemporary fonts, visit www.philsfonts.com/phils/sections/links1.html. (adj.) A *font family* is a complete set of fonts for the same typeface (usually includes normal weight, bold, italic, and bold italic fonts). *Font cartridge, font manager, font metric, font style, font utility, font weight.* See also **True-Type, PostScript**.

forward slash (/) See **slash**.

free-commerce (n.) Derived from *e-commerce, free-commerce* is an *e-commerce* service that is, at least initially, free to businesses who wish to sell products and services over the Web. (adj.) *Free-commerce host, free-commerce services.*

front end (n.) The user interface to a **back-end** server or program. Often a Web-based form will serve as the *front end* to a database. (adj.) Hyphenate the adjective: *front-end application, front-end infrastructure.*

FTP *File Transfer Protocol,* pronounced "F-T-P." (n.) An Internet-based method for transferring information from one computer to another. (v.) To **upload** or **download** files to/from another computer using File Transfer Protocol: *to FTP files, FTPed, FTPing.* (adj.) *An FTP site.* Use initial caps for File Transfer Protocol and other protocol names. When writing an *FTP address* in text, include the *ftp://* prefix, as in *my printer's FTP site, ftp://ftp.tradesvc.com/tsp/incoming/.* See also **HTTP**.

gateway (n.) A *gateway,* in the new-media sense, refers to the hardware and software setup that bridges the gap between two or more dissimilar protocols. The *gateway* both transfers information and converts it to a form compatible with the receiving system or network. For example, a *gateway* translates e-mail that comes from a proprietary system like America Online to an Internet e-mail format. The word *gateway* is often used imprecisely to mean a system that provides access, as in "CompuServe provides its users with a gateway to the Internet."

GB The abbreviation for *gigabyte*—one billion bytes.

GHz The abbreviation for *gigahertz*—one billion hertz. See also **Hz** and **MHz**.

GIS *geographic information system*

Gopher (n.) The Internet utility *Gopher* is most often spelled with an initial cap. It is a proper name like the World Wide Web. (adj.) *Gopher server, Gopher site, Gopher client, Gopherspace.*

Greenwich Electronic Time (GeT) A proposed standard for global Internet and e-commerce timekeeping. Based on Greenwich Mean Time (GMT), *GeT* is one of several initiatives intended to streamline communications and e-business by putting all Internet communicators on the same 24-hour clock. This standard has not received worldwide adoption, although large companies such as Microsoft and DHL have endorsed it, along with the British government.

GUI *graphical user interface,* pronounced "gooey."

hacker (n.) *Hacker, hacking.* A *hacker* is an individual who thrives on solving problems related to computer systems, programming, or software applications. *Hacker* usually refers to an expert, ingenious, nonconventional computer user who breaks the rules. A *hacker* may break into computer security systems as a part of problem solving and sometimes just to prove that it can be done. The tendency in the popular press is to use *hacker* generically to refer to the activity of both *hackers* and **crackers**. The difference is mainly that *crackers* operate with malicious intent. A person who breaks into a Web site with intent to cause damage is a *cracker*. The *hacker ethic* denigrates destructive use of computer expertise. (v.) *To hack.* See also **phreak.**

handheld (n.) A *handheld* usually refers to a **PDA** or computer that is small enough to fit in one hand. A *handheld* usually performs a number of functions and receives input from a keypad or device such as a pointer or barcode scanner. (adj.) *Handheld PC, handheld computer.* A **palmtop** is a type of *handheld device.*

hardcopy (n., adj.) Style it as one word. *Hardcopy* refers to the physical, paper version of a document or a printout of data as opposed to the electronic version, as in "I need the *hardcopy* as well as the electronic version." The opposite, not as common **softcopy** refers to an electronic version.

HDML *Handheld Device Markup Language*

HDTV *high-definition television*

help (n.) Online instructions on how to use a software application: *online help, context-sensitive help* (adj.) *Help menu, help desk, help line.*

hertz See **Hz.**

high-resolution See **resolution.**

Quick Reference List of Troublesome Terms

high-tech (adj.) Hyphenated as an adjective when it precedes the noun, and we propose it hyphenated as a unit modifier, too. Occasionally but not preferably seen as *hi-tech*.

hit (n.) A *hit* is an often-misunderstood Web site traffic statistic. Often the number of *hits* is taken to mean unique visitors or **pageviews**. However, a hit is the total number of files downloaded or accessed. One page may consist of several files, as the viewer of a page has to access not only the page itself, but also the graphics and any scripts running on that page. Therefore, one *pageview* may generate dozens of *hits*. This statistic is misleading when it is used to quantify a site's total traffic. A site receiving 1 million *hits* per month, if it is graphics-intensive, may receive fewer unique visitors with fewer pageviews than another site that delivers fewer files. See **Web traffic and advertising terms.**

HPGL *Hewlett-Packard Graphics Language*

home page (n.) The *home page* is the main page for a Web site. Spelled as two words, consistent with **Web page**. Every Web site has a home page; complex Web sites may have multiple *entry* pages, say, to feature specific products, but only one *home page*. The *home page* does not automatically mean the *opening screen*. Some sites use **splash screens** before the home page. An alternative, less common spelling is *homepage*.

horizontal portal See **portal.**

host (n.) On a network, a *host* is a computer that provides information or services to other computers. A *Web site host* is a computer that houses a Web site and makes it accessible to the Internet. (v.) To provide the hardware, software, and communications necessary for a computer network or Web site: "A local Internet service provider *hosts* our Web site."

hot plug, hot swap (n.) *Hot plugging, hot swapping.* The ability to add or exchange system components or peripherals on a computer while the power is on, and have the operating system recognize the device without restarting the computer. *Hot swap* is more often used for system components such as power supplies or hard drives, especially in connection with a server or RAID device. *Hot plug* is more often used in reference to peripheral devices connected to a computer by a **USB** or **IEEE 1394** connection. (adj.) *Hot pluggable, hot swappable,* also *hot plug, hot swap.* (v.) *To hot plug, to hot swap.*

hot swap See **hot plug.**

HTML *Hypertext Markup Language*

HTTP *Hypertext Transfer Protocol.* (n.) HTTP is the basic **protocol** or standard that allows documents to be delivered across the Web. Any **Web browser** can communicate with a server using *HTTP.* The *http://* prefix in a Web address tells the Web browser that the document conforms to the *HTTP standard.* If the communications protocol is not specified, the browser assumes that it conforms to *HTTP.* Capitalize the acronym and use initial caps when spelling it out. See also Web addresses in Keeping Up with Style; **FTP;** and **URL.**

hyper- words Most words using the prefix *hyper-* are written as closed compounds: *hyperlink, hypermedia, hyperspace, hypertext, hyperware,* but *hyper-reality.*

Hz *hertz. Hz* is a measure of the frequency of vibration (cycles per second) of an electromagnetic wave. One *hertz* is equal to one cycle per second. *Hertz* is capitalized by some; it's named for a German, Heinrich Hertz.

i- words *Jargon.* The *i-* prefix stands for Internet. *I-*words are spelled as hyphenated compounds (*i-content, i-publishing*) and also as closed compounds with an intercap to avoid pronunciation confusion *(iVillage.com, iPublish.com).*

ICANN *Internet Corporation for Assigned Names and Numbers*

IEEE 1394 (n.) The Institute of Electrical and Electronics Engineers standard number 1394 for a high-performance serial bus is *IEEE 1394.* (adj.) *IEEE 1394 technology, IEEE 1394 digital interface, IEEE serial bus.* See also **FireWire.**

impression See **Web traffic and advertising terms.**

imagemap (n.) A graphic image that has been divided into hyperlinked sections. An example of an *imagemap* is a national parks locator Web site featuring a map of the United States that allows users to click on a state to go to information on national parks in that state. Types of *imagemaps* include *client-side imagemaps* and *server-side imagemaps.* (v.) *To imagemap a graphic.* (adj.) *Imagemap software, imagemap editor.*

impression (n.) An *impression,* in Web terms, refers to one viewing of an online ad. An ad that receives 10 *impressions* may have been viewed by one person 10 times, or once by each of 10 people. See also **Web traffic and advertising terms.**

index (n.) In information technology (IT) terms, an *index* is a method for organizing and finding data in a database or table. An *index* is based on key fields or key words that identify records. *Programming index, data index.* On a Web site, an *index* is a hierarchical directory containing hyperlinks to other Web pages. (v.) *To index a Web site.*

index page (adj.) An *index page* is a directory page on a Web site that contains a hierarchical directory of links to other pages either on or outside the site. Large sites have many *index pages.* Smaller sites may use the **home page** as an *index* to all other pages on the site. See also **search engine.**

inhouse (adj.) and (adv.) One word. *Inhouse* work is performed within the company or organization as opposed to work that is **outsourced,** which is the opposite term (not *outhouse).*

inkjet (n.) *Inkjet* is a generic term for a type of printer that forms letters and images by spraying ink on paper. (adj.) *Inkjet printer, inkjet technology.* **DeskJet** and **Bubble Jet** are types of *inkjet printers* made by Hewlett-Packard and Canon, respectively. A **LaserJet** is a laser printer. See also **laser printer.**

input device (n.) *input device* is a generic term referring to any device used to enter data into a computer. The term encompasses the **mouse**, trackball, stylus, pointing stick, touchpad, graphics tablet, joystick, touch screen, keyboard, and scanner. The term *input device,* though stilted, is an option when it isn't clear what type of device a reader uses (mouse or touchpad, for example). Other generic terms, less inclusive, are *tracking device* and *pointing device.*

instant message (IM) (n.) An *instant message* is a message sent from one computer to another over a network or the Internet that appears in real time. The key distinction between *e-mail* and *instant messaging* is that *IMs* appear onscreen on the recipient's computer as soon as they are received; *e-mail* messages sit on a server until the recipient retrieves them. The word in its different forms is used as a noun, an adjective, and a verb. (n.) The message itself is an *instant message* or *IM* (pronounced "I-M"). The system is *instant messaging*. (adj.) *Instant messaging program, instant messaging service, instant messaging system.* Popular *instant messaging programs* include ICQ, AOL Instant Messenger, and Yahoo! Pager. (v.) *Jargon.* The acronym is often used informally as a verb: *to IM. We IM'd back and forth all afternoon."* IM, IMing, IM'd. Also called *Internet chat*.

internationalization See **localization**.

Internet (n.) When referring to the worldwide network of computer networks that communicate via **TCP/IP,** the word *Internet* is always capitalized: the Internet. It is a proper noun. (adj.) *Internet-based toy sales, Internet address.* The *World Wide Web* is just one part of the *Internet,* which also includes functions such as **e-mail, FTP, Gopher, IRC,** and **Telnet**.

intranet (n., adj.) Refers to a company's internal network of HTML pages. Do not shorten to *Net* or *net,* as this can be mistaken for a reference to *the Internet.* Refer to it as the corporation's *intranet site* (or *the intranet* on internal company documents). Do not refer to an intranet as a *Web* or *web.* Although they use the same technology, intranet sites are not *on* the Internet but rather are accessible from the company's **LAN** or **WAN**.

IP address *Internet Protocol address.* (n.) A computer's numeric address on a **TCP/IP** network. Because the Internet is one vast TCP/IP network, all Web sites have *IP addresses* that are assigned by InterNIC. *IP addresses* are written as four sets of numbers separated by periods, such as 207.67.134.127.

IRC *Internet Relay Chat*

ISDN *integrated services digital network*

ISP *Internet service provider*

IT *information technology.* (n., adj.) This acronym is used loosely to refer to any and all tasks, skills, and activities associated with computers, the Web, and new media. The people in *IT departments* purchase, install, upgrade, and troubleshoot software, hardware, peripherals, and related electronic systems. *IT* may mean anything from database management and programming to daily maintenance of a network, applications support, and Web site development. The oldest synonym for *IT* is *MIS* (*management information systems*), which gradually gave way to the less-clunky *IS* (*information systems*). Now the broader term *IT* is most frequently seen. It seems a boundless array of powerful, complementary technologies, which have forced even the stodgiest publisher onto the Internet, simply overtook the more static "system" model.

Java, JavaScript *Java* and *JavaScript* are often used (erroneously) as interchangeable terms. It is important to be aware of the difference. *Java* is a full programming language created by Sun Microsystems that was based on C^{++}. *JavaScript* is a simpler scripting language created by Netscape, more reminiscent of Perl or Visual Basic than C^{++}.

Both *Java* and *JavaScript* are used to create special effects for Web sites. However, *Java* is more likely to be the programming language behind a heavy-duty e-commerce development, while *JavaScript* is used to create smaller, "light duty" Web site effects such as **mouseover** animations.

just-in-time (JIT) (n.) Just-in-time is hyphenated when it refers to the manufacturing and inventory management strategy. (adj.) *Just-in-time production.*

key- words Compounds with the word *key* are variously open and closed: *keyboard, keycap, key click, key command, key data, key driven, key entry, key field, key frame, key in (v.), Keypad, keypal, keypunch, keystroke, keyword.*

killer app See **application.**

LAN *local area network.* (n.) A *LAN* is a group of resources including computers, printers, modems, and so on that are networked together. The computers are usually located within the same office or building. See also **Ethernet, WAN.**

laptop (n., adj.) A portable computer. Spelled as one word.

laser printer (n.) *Laser printer* is the generic term for a type of printer that uses laser technology to fix images on paper. A **LaserJet** is a type of laser printer.

LaserJet (n.) A *LaserJet* is a type of **laser printer** made by Hewlett-Packard, although it is often confused as a synonym for laser printer. The name *LaserJet* is a registered trademark of HP.

LISTSERV (adj.) *LISTSERV* is a registered trademark for the electronic mailing list management software marketed by L-Soft. Be careful not to misuse *LISTSERV* as a synonym for an electronic mailing list—even if the list is indeed managed by *LISTSERV* software. Wrong: *a LISTSERV list for pet fanciers*. (n.) As *LISTSERV* is a trademark, do not use it as a noun to refer to the electronic mailing list itself or in a generic sense to refer to any e-mail list based on another mailing list management software. The correct noun to refer to such a list—whether it is based on *LISTSERV* software or not—is an *electronic mailing list* or an *e-mail list*.

localization (n.) One of the major topics in software and Web development today is *localization*—customizing a software application or a Web site's content and features for a "local" audience. *Localization* goes beyond straight translation to include accommodation for local culture and beliefs. For example, Berlitz explains that when localizing Web graphics, "a person outside of the U.S. may not understand the relationship between the graphic of a pink piggy bank and your company's savings plan." (v.) *To localize, localized, localizing*. (adj.) *Localization industry*. Also called *internationalization*.

log on, log off (v.), **logon** (adj.), **login, logon** (n.) To *log on* to a network means to connect with the network by submitting the proper user name and/or password to gain access. *Log in* is an alternative for *log on*. To disconnect from the network, one *logs off* or *logs out*. Microsoft recommends against using *log in*; however, in common use both *log on* and *log in* are acceptable. The *Wired Style Guide* offers the advice that "*log in* and *log out* are more common in the Unix world." We advise using the terms consistently: Pair *log on* and *log in* with *log off*, and don't switch randomly among the variants in a publication or on a Web site. Use *log on to*, not *log onto*: *The user logged on to the network*. Usage note: It's important to remember that the *logon process* involves authorization. Don't use it for an open access system—don't say *Log on to our Web page*. (adj.) *Logon* can be used as an adjective: *logon password, logon routine*. (n.) *Login*. The account name an individual uses to connect to a network, Web service, **BBS**, and so on: *My login name is jsmith*. Also called *logon*.

low-resolution See **resolution**.

media (n.) 1. *Media* is technically a plural term (a count noun), with *medium* as its singular. *Media* as a singular term was once understood to refer only to newspapers. But *media* is now widely used as a collective noun to mean all forms of information dissemination including newspapers, magazines, television, radio, and the Web. That's what *mass media* means, and *media* has become shorthand for that collective (singular) sense. We recommend using the singular in this sentence: *Is the soundbite-hungry media to blame for the negative tenor of political dialog?* 2. *Media* are objects that hold computer data for purposes of backup or distribution. This category of *media* includes diskettes (floppy **disks**), **CD-ROMs**, hard disks, and magnetic tapes. 3. In a new-media context, *media* refers to the specific electronic techniques used to present content and can include text, audio, video, graphic enhancement, and animation. The term *mixed media* incorporates several different forms of media. 4. In a computer network, *transmission media* link individual workstations together. Types of *transmission media* include **coaxial** cable, twisted-pair wire, and **fiber-optic** cable. (adj.) *media feeding frenzy.* See also **multimedia** and *rich media* under **rich.**

megahertz See **MHz.**

meta tag (n.) *Meta tags* are words that describe the content of a Web page so that it can be indexed by a **search engine.** A casual visitor viewing a page with a Web browser does not see the *meta tags.* While most words with the *meta-* prefix are closed compounds, we spell *meta tag* as an open compound for clarity and because the preponderance of sources spell it that way. See also **Web browser, index.**

meta- words The prefix *meta-* means "about" and is usually written as a closed compound with the word it modifies. A *metalanguage* is a language used to describe other languages. A *metafile* is a file that contains other files. The fact that *The Metadata Company* is a registered trademark does not govern the style of the common noun *metadata,* meaning "data about data." The closed form is more frequently used, though *meta-data* is also seen. See also **meta tag.**

MHz *megahertz.* One MHz represents one million cycles or **hertz** per second. The clock speed of computer processors is measured in *MHz*—although computers are being released with clock speeds measured in gigahertz (GHz) or billions of cycles per second.

MIME *Multipurpose Internet Mail Extensions*

MIPS *millions of instructions per second.* It is never correct to use *MIP* or *MIPs* because MIPS itself is already plural and may be used only as a plural.

MIS *Management Information System.* A more common expression is IS (information system).

mixed media (n., adj.) Hyphenate the adjective: *mixed-media* show. See **media.**

MMX (adj.) See **Pentium.**

mouse (n.) A *mouse* is an **input device** that controls the movement of the cursor on the computer screen. That's easy. But what is the plural of mouse? The most popular variants are *mice* and *mouses.* We favor *mouses* as the plural for *mouse.* Although the *mouse* was originally named for its rough resemblance to the furry creature of the same name, there is very little association between the technical gadget and the rodent. A stiffer, more formal variant is *mouse devices.* (v.) Do not use *mouse* as a verb (*to mouse*); instead say, *Using the mouse, double-click* or *Using the mouse, point to.* The Free Online Dictionary of Computing (http://foldoc.doc.ic.ac.uk/) gives us the following table defining the five most common functions performed with the mouse and its buttons.

point	To place the pointer over an onscreen item
click	To press and release a mouse button
double-click	To press and release a mouse button twice in rapid succession
right-click	To quickly press and release the right mouse button
drag	To hold down the mouse button while moving the mouse

Click usually implies a click with the left mouse button or left-click, hence the use of *click* and *right-click.* However, if the audience of your publication has a beginner-level understanding of computers, do not assume that the audience understands; explain these terms. Note that not everyone uses a *mouse* as **an input device.** People also use trackball, stylus, pointing stick, and touchpad. *Input device, tracking device,* and *pointing device* are all generic (though stilted) alternatives for *trackball* and *mouse.*

mouseover *Jargon.* Closed compound. (n.) The **JavaScript** technique that allows a Web page element (usually a graphic) to change as the mouse passes over it. (adj.) *Mouseover effects.* Also spelled *mouse-over.* Sometimes called a *rollover.*

MP3 (n.) *MP3* is the file extension for a file compressed by the MPEG-1 Audio Layer 3 standard. This is one of the most common formats for transferring music over the Internet. What constitutes legal use and distribution of *MP3 files* is the hottest, most controversial subject in the music industry today, both online and off.

multi- words The prefix *multi-* usually adds a connotation of "many," "more than one," or "many times over." Words beginning with the prefix multi- are usually closed compounds: *multicast, multidimensional, multifrequency, multilaunch, multilayer, multilevel, multilingual, multimedia, multimode, multiplexing, multiplexor, multipoint, multiported, multitask, multithreading, multiuser, multivendor.*

multimedia (n.) *Multimedia*, used rather loosely to mean "a category of content," "a format," "a means of publication", or all three, includes text, graphics, audio, and video. (adj.) *Multimedia computer, multimedia software. Multimedia software* enables us to create online presentations that incorporate video clips, sound, and animation. See also **media.**

network (n.) A *network* is a group of interconnected computer systems and peripherals. When referring to an individual computer, it is *on* the network, not *in* the network. In the IT sense, *networking* refers to linking computers together on a network *(a networking specialist);* of course, it also refers to people-to-people interchange at meetings. (adj.) *Network adapter, network administrator.* (v.) *Network* is slightly more controversial when used as a verb. It is common practice to see network used as a verb—for example, *Network your computers for more efficient file sharing. Microsoft Manual of Style for Technical Publications* advises against the use of network as a verb. To be technically correct, their advice is that "computers are linked or connected, not *networked."* See also **LAN, WAN.**

newsgroup (n.) A *newsgroup* is an online discussion group that is distributed through Usenet, using the Internet protocol Network News Transfer Protocol (**NNTP**). The Internet is host to thousands of newsgroups. Only online discussion groups that are a part of Usenet are considered newsgroups. Non-Usenet online forums have other names, like *chat rooms* or *Web-based discussions.*

NNTP *Network News Transfer Protocol*

OEM *original equipment manufacturer.* (n., adj.) 1. *An OEM* is a company that manufactures generic hardware components such as monitors or **CD-ROM** drives to be marketed under other companies' brand names. *An OEM* is also a company that buys hardware or software components either to resell under its own brand name or to package and resell as part of a customized computer system. (v.) *OEM* is also used as a verb, consistent with both of the above definitions: *This monitor is OEM'd by ABC Corporation* (this monitor is originally manufactured by ABC Corporation). *We OEM XYZ Company's software as a part of our product* (we include XYZ Company's software as a part of our product).

The *OEM* reseller is also known as a *value-added reseller (VAR).*

offline (adv.) Disconnected and/or unable to receive data: *The printer is offline. Offline* has taken on the informal connotation of being away from one's computers and the Internet and unreachable by e-mail: *I will be offline while I am on vacation in the Amazon.* (adj.) *Jargon. Offline* can mean noncomputerized or not Net-connected—*the offline world* as opposed to *the* **online** *world.*

on the fly (adv.) **Dynamic** Web pages are composed of components collected and served on demand, or *on the fly.* (adj.) Hyphenate the adjective: *"On-the-fly editing* of video is editing live or without stopping the tapes" (Medialink Broadcasting Glossary).

online (adv.) and (adj.) *Online* has many meanings. We recommend using *online* as a closed compound for all of them. *Online* as an adverb can mean "in a computerized format" (versus print): *the tax forms are available both in print and online. Online* may mean "on the Web," "on **CD-ROM**," or "in electronic book format." The key is being available digitally as opposed to existing as a flat, printed document. If using *online* to mean "digital," be specific as to the medium; *online* has begun to imply *Web,* as in *to do research online.*

"To sign on to the Internet or another computer network": *I need to go online to check that fact.*

"To have Internet access available": *Vista's middle schools are all online.*

(adj.) In its adjective form, *online* means connected with the Internet or a computerized system: *online merchants, online order form, the online world.*

Online also refers to electronic equipment being connected and ready to receive data: the printer is *online*. The opposite is **offline**.

online community (n.) Either the entire population of Web-users or a specific niche community formed around a specific **portal**, Web site, chat room, e-mail list, **newsgroup,** etc.

outsource (v.) To use services or resources provided by a consultant or supplier outside a firm. Closed compound.

PageMaker (adj.) Adobe *PageMaker software* is used for desktop publishing. Adobe, like all publishers holding trademarks, prefers its trademarks to be used as adjectives followed by a generic term (such as *software)*, not as nouns or verbs. In reality, technical people don't follow that preference religiously, and neither do civilians.

pageview (n.) A pageview is a Web advertising term that refers to a Web site visitor viewing a page. If a visitor clicks on 12 pages on a Web site, this constitutes 12 *pageviews*. Online advertising rates are often based on the *pageview statistic*. See also **Web traffic and advertising terms.**

palmtop (n., adj.) See **handheld.**

PDA *personal digital assistant.* (n.) A *PDA* is a tiny personal computer. Also known as a **handheld** or *handheld PDA.*

PDF *Portable Document Format.* (n.) A *PDF* is the type of file created by Adobe's Acrobat software. (adj.) *A PDF file.*

Pentium (adj.) *Pentium* is a registered trademark of Intel, which prefers that it be used as an adjective accompanied by an appropriate noun, such as *chip, microprocessor(s),* or *processor(s)*. Related trademarks of Intel are *Celeron, Xeon, MMX, Pentium II, Pentium III, Pentium III Xeon, Pentium Pro,* and *Pentium OverDrive.* The *Pentium hardware* in its many variations is the successor to the Intel 486 chip; prior chips are the 286 and 386. The word *Pentium* should not be used to refer to chips such as the K6 that have comparable features but are not made by Intel. For Intel's trademark instructions, see www.intel.com/intel/legal/tmusage2.htm.

Photoshop (adj.) Adobe *Photoshop* software is used for image editing. Adobe prefers that its trademarks be used as adjectives followed by a generic term (such as *software).* Correct: *Color-correct these images using Photoshop software.* Incorrect: *We need to photoshop these images.* But we all use it that way.

phreak (n.) *phreaker, phreaking.* A person who specializes in breaking into telephone networks is known as a *phreaker* or a *phone phreak.* Phreaking often depends on theft of phone codes and is rarely a legal activity. (v.) *To phreak.*

plug and play (n., adj.) *Plug and Play (PnP)* is a standard developed by Intel. Computer systems that conform to this standard should automatically recognize and configure new hardware and peripheral devices. When referring to the standard itself, the noun should be capitalized. Technically, any reference to a device or operating system compatible with this standard should also be capitalized *(Plug and Play).* However, in common usage, the noun is more frequently written lowercase, sometimes hyphenated as *plug-and-play.*

Always hyphenate the adjectival compound: *a plug-and-play device, a plug-and-play computer,* and, though painful, *a non-plug-and-play device* (or revise to *not plug-and-play.*)

Jargon. *Plug-and-play* is also used (very) informally to describe a new employee who needs no training: "The help desk is a disaster; there's no time for training. We need a few recruits who are totally *plug-and-play.*"

plug-in (n.) A script, utility, or program that, when added to a larger software application, enhance the functionality of that application. Netscape Navigator, QuarkXPress, and Adobe Photoshop are programs that make extensive use of *plug-ins.*

point-and-click (adj.) Always hyphenated: Both Microsoft Windows and the Mac OS feature a *point-and-click interface* that allows a user to access data and activate programs by selecting a file name or icon with a mouse or other input device and clicking on it.

pointing device See **input device.**

POP *Post Office Protocol,* a protocol for receiving e-mail from a remote server (the most recent version of this protocol is POP3). Also a local dial-up number for an Internet service provider (sometimes spelled *PoP).*

port (n.) Ports are external sockets or connectors on a computer to which peripheral devices are attached: *a data port, serial port, parallel port, COM port, modem port.* A *port* is also a software construct. In a communications network such as a **TCP/IP** network, a port is the endpoint to a logical connection that is identified by a unique *port number*. (v.) *To port to/to port from*. To convert data or software from one computer platform to another, as from Macintosh to PC format. Software usually has to be rewritten to be usable on a different platform. (adj.) The adjectival form is *portable,* meaning "capable of being ported." Highly *portable programs* are easily *ported*.

portal (n.) A *portal* is a Web site that offers a jumping-off point to the Web. *Portals* offer concentrated resources, news, and services to attract and hold repeat visitors. They often offer personalization features that allow registered users to customize the data that the portal will **push** when they sign-in. A *vertical portal* concentrates on a tightly focused niche market, such as engineers or desktop publishers. A *vertical portal* for mechanical engineers is iCrank.com; it offers features such as engineering reference data, links to vendors, software and hardware, technical discussion forums, and news of upcoming conferences. A *horizontal portal* offers a broad range of services and features to appeal to the broadest possible range of people. Major *horizontal portals* such as Excite and Yahoo! offer myriad features, including personalization, free e-mail, a search engine, news, weather, chat rooms, stock quotes, shopping, and horoscopes. Acceptable references are *portal* and *Web portal*. Do not use *portal site*.

post, POST (n.) A *post* is a message sent to an Internet newsgroup, bulletin board, or electronic mailing list. A *post* can be read by all subscribers to a newsgroup. A *POST* (all caps) is a *Power On Self-Test,* the self-test a computer performs when it starts up. (v.) *To post* is to send a message to an Internet newsgroup. See also **cross-post.**

PostScript (n., adj.) *PostScript* is a programming language created by Adobe. It describes the appearance of a page in terms of fonts, graphics, text, and layout to ensure exact replication in printing, whether it is to a professional imagesetter or a personal *PostScript-compatible* laser printer. A registered trademark of Adobe. *PostScript printing technology, language, page-description language, interpreter, font, printer,* and *RIP.* Use *PostScript* or *PostScript-compatible* to describe a printer only if it is *Adobe PostScript-compatible.*

POTS *plain old telephone service* or *plain old telephone system*. This is standard telephone service, in contrast to the newer systems based on high-speed, digital communications lines such as **ISDN**. The term is sometimes written with initial caps when it is spelled out: *Plain Old Telephone Service.*

ppi A graphic design acronym for *pixels per inch* or *points per inch*, it's also a printing industry acronym for *pages per inch.*

ppm *pages per minute*

preflight *Jargon.* (adj.) Always a closed compound. The thorough review of an electronic publication before it is sent to the printer for output: *preflight check.* (v.) *to preflight*: *Preflight the annual report files before they go to the printer.*

print-on-demand (n.) *Print-on-demand* is a relatively new process for printing books efficiently and cost-effectively one copy at a time. (adj.) Barnes & Noble struck an agreement with IBM to provide *print-on-demand equipment* for its distribution centers.

program (v.) To write programs. (n.) A *program* is a set of instructions that tells the computer what to do. *Software* is the common term for an executable program: *a word processing program* or *word processing software.* (adj.) Programming language.

protocol (n.) An agreed-upon set of rules or standards that enable disparate computers or systems to communicate with each other. Common *protocols* are **HTTP, FTP, POP,** and **TCP/IP**. *Protocol names* are usually written with initial caps when they are spelled out: *Hypertext Transfer Protocol.* (adj.) *protocol port, protocol stack, protocol suite.*

pull (adj., v.) The common form of information access across the Web or a computer network, *pull technology* delivers information or data only when a user directly requests it. Clicking on a hyperlink or requesting a file is a way that a user *pulls* information from a server. Contrast with *push technology.* See also **push.**

pure play (adj.) A *pure play* company has its entire operation online. A *pure play* Internet company does not have a **bricks-and-mortar** presence. Also called a *pure Internet company.*

push (adj.) *Push technology* delivers a program of news and information to an individual's computer or handheld device. (v.) Information is delivered or *pushed* to the user either on a preset schedule or upon request. The user specifies what information he or she wants delivered (stock quotes, news headlines, local weather, entertainment news, etc.) and how often. Web **portals** often feature personal **home pages** that *push* custom programs of news and information to users who sign up for the service. Contrast with **pull** technology.

qwerty (adj.) A *qwerty* keyboard is the standard typewriter and personal computer keyboard, identified by the first six letters on the top left of the keyboard. Also spelled *QWERTY*. The *qwerty keyboard* was originally designed to slow typists enough so that the keys of a manual typewriter would not jam. The **Dvorak keyboard**, developed to facilitate faster typing, has not caught on, but try the concerto for violin and orchestra in A minor, Opus 53.

radio button (n.) Two words. *Radio buttons* are used on Web pages and online forms when the person filling out the form must choose one of several options. Like a pre-digital car radio that had big buttons to push to select a station, *radio buttons* can be selected only one at a time.

RAM *random access memory*

real time (n., adj.) Something that is telecast, broadcast, or relayed when it is actually happening, as opposed to a recorded or delayed transmission event. *Millions of viewers tuned into the NetAid concert that was broadcast over the Internet in real time.* (adj.) Hyphenated. *Real-time animation, real-time clock, real-time compression, real-time conferencing, real-time image, real-time operating system, real-time video, real-time information system.*

repurpose (v.) To reuse content and/or electronic coding/layout for multiple applications such as several types of print publications, Web pages, and CD-ROMs.

resolution, high resolution, low resolution (n.) In graphic arts and computing, *resolution* refers to the fineness of detail and clarity of an image. The term is used with bitmapped graphic images, printers, scanners, and monitors. The more dots or pixels per square inch, the higher the *resolution* and quality. *Low resolution* and *high resolution*, when used as nouns, are open compounds. (adj.) When used as adjectives, the compounds are hyphenated: *a high-resolution image, a low-resolution image.* Jargon: *high-res* or *hi-res, low-res.*

reverse auction See **buyer pool.**

rich *Rich* adds the connotation of "enhanced," usually with **multimedia** effects. *Rich e-mail* (n.) incorporates features in addition to plain text. Popular enhancements include sound and graphics. *Rich-media* (n.) is online content that is not static but incorporates interactive devices such as fill-in forms, sound, and animation.

right-click Two words, hyphenated. See **mouse.**

rollover See **mouseover.**

ROM *read-only memory*

scalable (adv.) Able to be expanded and reduced to the size and scope necessary. *This software is fully scalable; it can serve anyone from home-based Web creators to design teams at major corporate sites.* (v.) *to scale up* and *to scale down* (preferred to *upsize* and *downsize*).

SCSI *Small Computer System Interface.* Pronounced "scuzzy."

search engine (n.) A *search engine* is any program that allows users to locate specific data on the Web, a computer, a network, or in a database based on a set of parameters that the user specifies. Once a parameter (**keyword**) is entered, the search engine locates all possible matches in its database. *Search engines* are frequently associated with **Web search sites** such as Excite and AltaVista, which feature powerful proprietary search engine technology. These sites use **spiders** to query the Internet, collecting data on Web pages that will be indexed and incorporated into a searchable database. *Search engines* are not just Web-based. Microsoft Windows Explorer's Find feature is an example of a common desktop *search engine.* See also **Web directory.**

secure (adj.) A secure Web server conforms to one of the major security protocols to prevent unauthorized or destructive access. When a server or an area of a Web site is *not* secure, refer to it as *nonsecure* (some use *insecure*, but we prefer a word with less of an emotional overtone).

server (n.) A *server* is a computer or device on a network that manages a specific type of resource: *Web server, application server, fax server, mail server, file server, network server, print server, database server, remote access* *server.* A *server* is shared by multiple users. (adj.) *Server application, server farm, server-side script, server-side.*

setup, set up, Setup (n.) The way a computer's hardware and software is configured is its *setup* (one word). *Setup* (initial cap) refers to the installation program that comes with most applications (setup.exe): *Run Setup to install the program on your computer.* (adj.) The adjective is spelled as one word: *setup routine.* (v.) *Set up* is two words when used as verb meaning to install or configure hardware or software.

SGML *Standard Generalized Markup Language.* SGML is a tagging or coding language of which **HTML** is a subset.

shock (v.) See **Shockwave**.

Shockwave (adj.) *Shockwave* files are created by Macromedia Director software. *Shockwave* technology has become the de facto standard for online multimedia productions—so much so that the word is becoming synonymous with *high-tech* multimedia presentations, whether they were produced with Director software or not. A file created in Director, then compressed for distribution by Web or CD/DVD, is *shocked: shocked file, shocked fonts.* See also **Flash**.

sign on (v.) *Sign on* is used as a synonym for **log on** when referring to connecting to a network. *Log on* is the preferred term. (adj.) *sign-on name.*

single-click Use *click* instead. See **mouse**.

slash (n.) *Slashes* (/), also called *forward slashes,* are used in Web addresses, such as http://www.eeicommunications.com/press/books.html. File names for IBM-compatible computers use the **backward slash** (\ ). *The file can be found on the network at y:\public\newsletter\file.doc.*

softcopy (n., adj.) Softcopy is the digital counterpart of **hardcopy,** which is printed material. So *softcopy* is an electronic file.

spam (v.) Use of the term *spam* has crossed over from the hacker's lexicon into mainstream English. Originally, the verb form meant to crash a program by inputting excessive amounts of data. It now generally refers to unsolicited, mass-mailed, and unwanted e-mail or newsgroup postings. (n.) In its noun form, *spam* refers to the actual message or messages— Internet junk mail.

Although the word *spam* is based on a proper adjective, we lowercase it. SPAM is the registered trademark of Hormel Foods Corporation. We found the following useful bit of advice on the spam.com Web site, which is owned by Hormel Foods:

"Use of the term 'SPAM' was adopted as a result of the Monty Python skit in which a group of Vikings sang a chorus of 'SPAM, SPAM, SPAM…' in an increasing crescendo, drowning out other conversation. Hence, the analogy applied because UCE (unsolicited commercial e-mail) was drowning out normal discourse on the Internet.

"We do not object to use of this slang term to describe UCE, although we do object to the use of our product image in association with that term. Also, if the term is to be used, it should be used in all lower-case letters to distinguish it from our trademark SPAM, which should be used with all uppercase letters." (www.spam.com/ci/ci_in.htm)

spec *Jargon.* (n.) As a noun, *spec(s)* is short for *specification(s),* as in *The specs for that print job are 20 copies, in Garamond type and double-spaced.* (v.) As a verb, *spec* is short for *specify* and means "to write instructions." Spelling of *spec* as a verb isn't pretty. Some dictionaries advise *speccing* and *specced* or *spec'd.* Because it's an informal word to begin with, we prefer the informal conjugation: *spec'ing* and *spec'd. Specked* is seen, but is not preferable. Working to *spec* means "doing it just as planned."

spider (n.) A *spider* is a program that collects data on Web pages to build **search engines.** The spider follows live links from Web site to Web site, retrieving data on the pages it encounters according to its search algorithms. *Spiders* are the agents that help build the databases behind search engines. (v.) *To spider, spidered, spidering.*

splash (adj.) A *splash screen* is an image that displays onscreen while a program is loading. The *splash screen* often displays the product name, logo, copyright data, and information about the author(s). A *splash page* is a flashy graphics and animation-intensive Web page presented to visitors before they reach the site's home page. Also called a *doorway page.*

startup (n.) and (adj.) Much like *voicemail,* which began life as an *E-What?* entry as *voice mail* but has moved to the solid form in widest use by summer of 2000, *startup* began life in newspapers like *The Washington Post* as *start-up* and is now most often seen closed up. A *startup* is a new company.

A *startup company* carries the connotation of being a new Internet company, perhaps because *Internet startup* is the usual usage. (v.) Use as two words: *To start up yet another online bookstore would be madness until the pioneers figure out how to break even.*

streaming (adj.) *Streaming* music and movies download in a continuous stream from the Web. *Streaming media, streaming video.*

SWOP *Specifications for Web Offset Publications*

systems analyst (n.) Not *system analyst.* A person who designs, develops, and modifies information systems. Plural: *systems analysts.* Also *systems engineer.* Contrast with the job descriptions that use singular terms: *system administrator, network administrator, application programmer, application developer.*

TCP/IP *Transmission Control Protocol/Internet Protocol*

techno- words Closed compounds: *technoanxiety, technobabble, technocrat, technofiend, technographer, technologitis, technophobe, technophile, technostress.*

tele- words Words beginning with the prefix *tele-* are proliferating. They are also always written as closed compounds. Some of the recent occurrences of *tele-* we have found: *telecommute, teleconference, telecommuting agreement, teleconference, teleworker, telework, telework coordinator, telework agreement, telework inventory, telework program.*

Telnet, telnet (n.) *Telnet* is a communications protocol that allows users to log on to a remote computer through a TCP/IP network such as the Internet. *Telnet* is capitalized when the word refers to the *Telnet* communications protocol itself. In Unix usage, the protocol may be spelled all caps— *TELNET.* (adj.) *The Telnet protocol, a Telnet command, Telnet extensions, Telnet options, a Telnet session.* When *telnet* refers to a program that implements *Telnet,* it may be lowercased: *a telnet program, a telnet client.* (v.) *Jargon.* The word is lowercased as a verb meaning "to access a remote computer using the *Telnet* protocol": *telnet, telnetted,* and *telnetting* (also spelled *telneted* and *telneting*).

touchpad One word. See also **input device.**

tracking device See **input device.**

TrueType (n.) *TrueType* is a scalable font technology developed jointly by Microsoft and Apple. *TrueType* is the predominant font technology today. (adj.) *TrueType fonts* are standard on all computers running the Windows operating system. The word is always spelled as a closed compound with two capital Ts.

typeface (n., adj.) A *typeface* is a design for a set of characters (letters, numbers, and common punctuation and symbols). Common *typefaces* are Arial, Helvetica, and Times New Roman. *Typefaces* fall into two broad categories: serif and sans serif. For a discussion on the difference between a font and a typeface, see **font**.

UNC *Uniform Naming Convention*

Unix (n., adj.) *Unix* is a popular multiuser, multitasking operating system originally developed by AT&T Bell Laboratories. Our recommendations on styling *Unix* with an initial cap come with a caveat. *Unix* was originally spelled all caps even though the word is not an acronym. Mainstream, popular, and nontechnical writing usually styles *Unix* with an initial cap only. Technical publications and several technical dictionaries spell *UNIX* with all caps. We feel the trend is toward initial cap only. Our advice is to use *Unix,* unless the specific public you're writing for expects *UNIX.*

upload (v.) To transmit a file from your computer to a remote location such as an FTP site. See also **download.**

URI *Uniform Resource Identifier*

URL *Uniform Resource Locator.* (n.) Usually pronounced "U-R-L," not "earl," so it's *a URL,* not *an URL.* A *URL* is a standard Web address. It consists of a Web protocol and a domain name. A longer *URL* may include one or more subdirectories and/or a file name.

The *URL* for EEI Press's book list is http://www.eeicommunications.com/press/books.html. The components of this URL break down as follows:

http://	protocol
www.eeicommunications.com/	domain name
press/	subdirectory
books.html	document name

The domain name portion of this *URL* is not case sensitive. Therefore it can be rendered without harm in several different ways—

www.EEIcommunications.com, www.EEICommunications.com, or www.eeicommunications.com. That can come in handy for printing the address out in a way people can remember. Sub-directory and file names *may be* case sensitive, so they should always be set exactly as specified in the file name. Therefore if the *URL* is http://www.eeicommunications.com/Press/Books.html, *Press* and *Books.html* should be capitalized. Web browsers may not find the file otherwise.

Because **HTTP** is the standard protocol associated with Web addresses, current usage is to drop the *http://* prefix when styling a URL in text. We need to write only www.eeicommunications.com rather than http://www.eeicommunications.com. The time to use the prefix is when the URL is for a site that uses a protocol other than HTTP (such as an **FTP** site).

Some publications have adopted the style of dropping the *www* from the URL when presented in text. This stylized approach assumes either that readers know to add the *www* and/or that many Web browsers (such as Netscape Communicator) automatically include the *www* prefix. *Working Woman* magazine uses this style consistently throughout, with *URLs* styled as follows: workingwoman.com, mayo.edu, and whitehouse.gov. (See Keeping Up with Style for further information on styling Web addresses in text.)

URN *Uniform Resource Name*

USB *Universal Serial Bus.* See also **FireWire.**

Usenet (n.) Network of newsgroups that operates across the Internet. Originally spelled *USENET.* (adj.) *Usenet groups.*

utility (n.) A program that performs a specific, narrowly defined function, usually relating to computer system maintenance. (adj.) *ScanDisk is a Windows-based utility program.* See also **application.**

vector graphic (n.) A *vector graphic* consists of a collection of lines rather than patterns of dots or pixels (as in **bitmapped** or raster graphics). Vector graphics are more **scalable** than bitmapped graphics.

vertical portal See **portal.**

video- words Words formed with the *video* prefix are usually, but not always, closed compounds: *videocam* (but *video camera*), *videoconference, videodisc, videogame, videotape* (but *video adapter* and *video port*).

viral marketing (n.) A marketing technique in which people are encouraged to contact friends to recommend the product.

virtual terms (adj.) Virtual is added to just about any word to add the sense of not real, as in existing in conceptual space, but not physical space. Terms including *virtual* are written as open compounds: *virtual office, virtual employees, virtual meeting, virtual reality, virtual desktop, virtual private network, virtual disk.*

voicemail (n.) *Voicemail* is used to refer to both the system *(All the employees have voicemail)* and, more informally, to the individual messages *(I left you several voicemails).* This term is a closed compound in most sources researched. (adj.) *Voicemail system, voicemail message.*

WAN *wide area network.* (n.) A **network** that connects computers that may be geographically distant, usually relying on telephone or satellite links. See also **LAN**.

WAP *Wireless Application Protocol*

-ware words (n.) *-ware* words are a takeoff on hardware/software and generally refer to classes of software products, such as *groupware, intraware,* and *middleware.* They usually appear as closed compounds, like their predecessor words. For example: *shareware, freeware, shovelware, vaporware, firmware, wetware, bioware.*

WBT Acronym for *Web-based training.* These are training programs, usually interactive, that are based on a Web site and operate over the Internet. See also **CBT** *(computer-based training).* WBT is also an acronym for *Windows-based terminal.*

Web directory See **directory.**

Web page (n., adj.) It's hard to say exactly what a *Web page* is—the term is used loosely—but non-tech people loosely understand what it means to say *New articles are added to The Editorial Eye Web page monthly.* A **home page** is not necessarily the opening page of a site; a so-called *Web page* can be many pages long (in terms of both screenfuls and sections), but it's not necessarily the same as a *Web site* (though, literally and practically speaking, a site is a collection of electronic pages formatted in HTML). Is that perfectly clear?

Two universities define the term this way for their site users (punctuation theirs):

Definitions of Web Page

A *web page* provides access to networked information and databases via a part of the Internet called the World Wide Web. The University of Delaware Library has a web "site" called a *home page.* The Library home page provides access to Library information, services and databases, including the Library Networked Databases. The Library Networked Databases page provides access to Library databases that are networked on the Internet. This way, you can access several Library databases from a single web page…

Web page—entry point in a World Wide Web information site; often called a home page. To create your own Web page, see Your Auburn University Web Page.

Perhaps this will help. EEI Communications has a Web *site,* made up of a collection of *Web pages,* one of which is the central *home page.* Each company division has its own *Web pages,* some of which are multilinked to outside pages and all of which are crosslinked to relevant services of other divisions. Each of these subdivisions has its own *home page,* as well. The whole schmear is a *Web site.* Some people take the shortcut of using *Web page* to mean "any or all of the content on a site," and as a synonym for *Web site,* as well. It would be clearer to use *Web page* to refer to a single document (corresponding to one specific URL) in a collection of content at a site. Don't hold your breath, though.

Web search site See **search engine.**

Web traffic and advertising terms The Web and e-commerce have given rise to a whole host of essential new statistical terms for measuring visitation or traffic on Web sites. These statistics help determine advertising rates, popularity, and attract funding for Web entrepreneurs. A **clickthrough** (n.) is a Web advertising term meaning a click by a viewer on an Internet banner ad, which triggers the associated hyperlink. (adj.) The clickthrough rate refers to the percentage of viewers of a Web banner ad who clicked on the ad. **CPM** refers to the cost to an advertiser per thousand impressions or pageviews. Web advertising rates are usually based on this figure. An **impression** in Web terms refers to one viewing of an online ad. An ad that receives 10 impressions may have been viewed by one person 10 times, or once by each of 10 people. A **pageview** is a Web advertising term that refers to a Web site visitor viewing a page. If a visitor clicks on 12 pages on a Web site, then this constitutes 12 pageviews. Online advertising rates are often based on the pageview statistic. **Hits** is an often-misunderstood Web site traffic statistic; see page 60. A unique *visitor* to a Web site is just that—one person, one visit.

Web words (n.) *Web* is capitalized when it stands alone, short for the proper noun *World Wide Web*. (adj.) *Web* is also capitalized when it forms part of an open compound, as in *Web technology, Web address, Web-based* ("Web-based merchants"), *Web-centric* ("a Web-centric commerce model"), *Web browser, Web page, Web portal, Web ring, Web site*. There are a few cases in which *Web* forms part of a closed compound; in these cases it is lowercase: *webmaster, webzine, webcast, webonomics*.

Web-based (adj.) An application or enterprise that is completely grounded in the Web.

webcast (n.) We style *webcast* as one word with a lowercase *w* in recognition of the word it is based on, *broadcast*. A *webcast* is a transmission of an event like a concert or conference over the Internet. The event may be live or recorded. (adj.) *Webcasting*. (v.) *To webcast*.

Web-enabled (adj.) An application or enterprise that makes use of the Web, but is not completely based on the Web.

webmaster *Webmaster* was originally a take-off on *postmaster*. Like *postmaster, webmaster* is written as a closed compound, and Web is not capitalized. The *webmaster* is the person responsible for building and maintaining a Web site.

WebTV (n., adj.) *WebTV* is a registered trademark of Microsoft. This product helps people surf the Web using their television and a set-top box. The name is often mistaken for a generic term. It's not. *WebTV* service can be enjoyed only by those who subscribe to Microsoft's WebTV Network.

webzine See **zine**.

Windows (adj., n.) The ubiquitous operating system registered as a trademark of the Microsoft Corporation. *Windows* comes in different flavors; common versions are *Windows 3.1, Windows 95, Windows 98, Windows 2000, Windows NT,* and *Windows CE.* A software program designed for the *Windows* operating system should be referred to as a *Windows-based program* rather than a *Windows program.* Microsoft's rules for using their trademark terms may be found at www.microsoft.com/trademarks. We all break them, all the time.

WML *Wireless Markup Language.* (n., adj.) WML is the markup language used by wireless devices. Allows formatted text to be displayed on a telephone microbrowser screen across a wireless network. Closely related to **HTML.**

workaround (n.) A *workaround* is the creative solution we develop when an existing system either doesn't work properly or doesn't cover all of the contingencies. (v.) Two words: *We need to work around this problem.*

workflow Closed compound. (n.) *Workflow* is an organizational concept, like **workgroup.** A *workgroup* is task-centric. In an organizational setting, a *workflow* consists of specific processes that need to be accomplished and who needs to do them in order to attain a specified outcome. (adj.) *Workflow processes, workflow automation.*

workgroup Closed compound. (n.) A *workgroup* refers to a group of computer users who share files. Also, on a network, a *workgroup* is a specific organizational unit. Belonging to a *workgroup* gives a user access to specific directories, applications, and files.

workstation (n.) A *workstation,* when referring to a computer, usually means a high-powered database and graphics-crunching personal computer. In a network context, workstation refers to an individual's personal computer, a network client as opposed to the **server.**

World Wide Web (n., adj.) Always preceded by the article *the.* Most commonly referred to as *the Web,* less commonly as *the WWW.* See **Web words.**

XHTML *Extensible Hypertext Markup Language.* A hybrid of **HTML** and **XML.**

XML *eXtensible Markup Language*

Yahoo! (n.) *Yahoo!* is a trademark of Yahoo! Inc. Along with *eBay* and **E*TRADE**, this name presents us with one of our least-favorite areas of e-style. Do we respect the company's style and include the *!* punctuation no matter where it falls in the sentence? Or do we drop the happy *!* and impose our style on the company's trademark? Our choice is to keep the *!* in *Yahoo!* See Keeping Up with Style, page 8, for a discussion of our decisionmaking process. (So why did we pick *E-What?* as a title? Just for a bit of mischief.)

zine *Jargon.* (n.) A *zine* is an informal newsletter/magazine. It usually appeals to a tightly defined niche and has slightly underground appeal. When a *zine* goes online, it becomes a **webzine** or **e-zine.** These digital zines have the same offbeat, niche appeal, but they reside on Web sites or are distributed by e-mail newsletter. We spell *webzine* as a closed compound with a lowercase *w* in recognition of the word's origin in the word *magazine.*

3. Planning Your Online Style Guide

FOR MANY EDITORS AND WRITERS, THERE IS NO OPTION: They must use the guide that has become the style bible for their field so their publications will compare favorably with those of colleagues and competitors.

Other communicators can pick a guide that offers the most advice for the editorial challenges and constraints they encounter, or a guide that reflects decisions and biases consonant with their own. (Corporate managers have been known to pick the Associated Press style manual purely because they do not "like" the series comma, even though this style manual is most useful to newspaper and newsletter journalists.)

The major style guides were created to be instructions for the editors and writers in a specialized publishing context. References are good, but don't just grab the first one you happen to see—choose one that relates as much as possible to your subject matter and audience. Most of us pick a main guide from among these comprehensive titles and then depart from it as necessary or as we wish to.

Government agencies are apt to reach for the *United States Government Printing Office Style Manual,* more commonly called *GPO Style* (but since the latest edition is 1984, it's ancient as style manuals go). If your organization deals with medical topics, the *American Medical Association Manual of Style* is an obvious choice. Many popular publications default to the

Associated Press Style Book and Libel Manual, which has the advantage of being updated frequently. The *Chicago Manual of Style* has a slight academic bias but covers many of the issues book publishers face. There are plenty of other specialized style guides, a few of which are listed here:

- *Publication Manual of the American Psychological Association*
- *Scientific Style and Format: The CBE* (Council of Biology Editors) *Manual for Authors, Editors, and Publishers*
- *The ACS* (American Chemical Society) *Style Guide: A Manual for Authors and Editors*
- *Mathematics into Type: Copy Editing and Proofreading of Mathematics*
- *Microsoft Manual of Style for Technical Publications*
- *MLA (Modern Language Association) Style Manual and Guide to Scholarly Publishing*
- *The New York Public Library Writer's Guide to Style and Usage*
- *Read Me First! A Style Guide for the Computer Industry* (based on *Sun Microsystems Editorial Style Guide*)

In addition to using a main style guide, many editors compile personalized stylesheets or circulate style memos from time to time to cover the issues particular to their work and to make sure everyone knows about changes and new decisions. The goal of all these sorts of guidelines—from major tome to memo—is to offer commonsense advice that doesn't require too many contortions of memory, and rules that don't invite a lot of exceptions.

WHAT'S OUR STYLE?

Style guides should be easy to understand and offer clear examples that apply to the work at hand, or people won't consult them, much less use them. Never underestimate the "working confused" factor: Even if it says "We use Chicago Press style," the typical organization has at least a few people asking each other specific style questions and never being entirely sure of the answers: Do we always use round bullets for lists? When do we use figures for numbers, and when do we spell numbers out? Should it be *Board of Directors* or *board of directors,* and *Chairman* or *Chair* or *chair*? We use the % sign with statistics in our ad copy, but we write out *percent* in our annual report. Which should it be?

These and innumerable other questions are matters of editorial style that are best addressed consistently by everyone. But the default action for a lot of people when a style squabble comes up is to fall back on what their favorite English teacher or pickiest boss required them to do ("Always put a comma before *and*"). Others want the freedom to change style to suit different situations. And still others, with less tolerance for inconsistency, want to call on authoritative references for what they hope will be more or less hard-and-fast rules.

If, after consulting the recommendations contained in *E-What?* and one or more of the major guides or a specialized guide in your field, there's *still* a lot of disagreement and uncertainty leading to inconsistencies and "errors" in your documents, realize that you may need to formally give people permission to do certain things differently. You may need a comprehensive handbook of your own style guidelines. This section is an overview of the main tasks involved in planning your own style guide, making it easy to use, and updating it.

A caveat at the outset for those who want things nailed down: Style isn't the same as grammar or usage. You won't find the "right" answers in any single reference. Style is always a matter of choice. Especially when you are creating a guide from scratch,

TAKING DISABILITIES INTO ACCOUNT

Are some of your style guide users visually impaired? Specialized software will allow them to hear text-based "alternate" messages (ALT tags, in HTML coding) that you can supply to explain navigational graphics and pictures. The Rehabilitation Act, amended in 1998 with Section 508, requires agencies to give federal employees with disabilities equal access to information, computers, and networks. How this mandate will apply to the Internet is still being worked out. In contrast to the Rehabilitation Act, the Americans with Disabilities Act (ADA) applies to agencies that do not receive federal aid. Guidelines for meeting compliance standards are still being written. More information is available from the Federal IT Accessibility Initiative site, **www.section508.gov/updateinfo.html**. You may also want to visit **www.w3.org/WAI** or **www.dinf.org/gsa/coca/law_pol.htm**.

you'll make countless large and small judgment calls about what style will work best for *your* editors and writers, *your* readers, and *your* publications.

CREATING A GUIDE

Whether you're the style guardian of your own work or the resident Style Keeper in a big shop, your aim is to make consistent calls using rationales that cover most cases. Guidelines that you formalize for yourself are *proprietary*—that is, they belong and apply solely to you. They can take the form of a simple alphabetical stylesheet or a comprehensive guide, and they can be in a notebook, on an intranet, or on an extranet so the world can access them. This section offers a description of considerations for both print and electronic guides.

You don't have to care what styles other companies follow once you've made your own thoughtful (or eccentric—hey, it's your jargon!) decisions. That's the first characteristic of having your own guide: It's not about your company adapting to a style but about adapting a style to fit your company. *You create it, so it can be highly specific to your needs.* It's going to reflect your preferences, trademarks, and other special matters dictated by the legal liabilities or branding goals inherent in your publishing efforts.

The second characteristic of having your own guide is this: It can reflect a consensus that will ensure that the guide gets used. *You create it for your colleagues, so*

> **TWO KINDS OF DIALOGUE WITH ONLINE STYLE GUIDE USERS**
>
> For the sake of both efficiency and courtesy, online style guide design should include clearly labeled links, icons, and buttons to make sure users always know where they are in the document. These tools tell them how to return to a previous page or the opening menu, and how to go forward on a specific path. Online style guides work best if someone is designated to receive questions and comments from users. Offer a feedback form or a simple e-mail link to the document's editor or webmaster—and acknowledge user messages (with a thank-you autoresponse, at the very least). Both the success of the guide now and getting users to help with the preparation of updates depend on both forms of dialogue.

they need to become involved in making sure it contains answers they need. Every organization has its own floating field of frequently asked questions—pinning them down and getting them listed is a good way to get started, and a good way to get people involved in the project. From the beginning, the guide should be perceived as a source of help, not just another time-consuming requirement. Ask everyone associated with writing, editing, and reviewing publications in your shop to fill out a short survey about what they have most trouble remembering the rules for—and ask for examples of iffy stuff from your own publications.

The third characteristic of having your own guide is that it can reinforce a professional image. You can offer the clients and consumers of your publishing products consistent quality. *You create it for your organization, so it should reflect careful consideration of editorial and design attributes widely acknowledged to affect credibility and easy accessibility.* Eliminating the inconsistencies makes it less likely that readers will be distracted from the main message of the document. Even though many readers won't recognize style inconsistencies as errors, they will be left with the impression that the document lacks polish.

Creating a stylesheet to cover the frequently encountered issues that require special handling in your publications may be enough to get you started on the road to a full-blown in-house guide. A style guide is a huge undertaking, but, fortunately, you don't have to invent the process. You can learn a lot from observing how major style guides are compiled—the magic words are *systematically* and *incrementally.* When it comes to the classic components of a style guide, you'll find they fall into these categories:

- **Word usage.** List and define commonly used acronyms and abbreviations. List do's and don'ts (e.g., "Refer to our organization as *the firm,* not *the company.*" "Refer to people who work for the firm as *associates,* not *employees.*"). Also list specific terms, such as the names of departments or publications. Frequently confused terms should be standardized: "Use *ensure* to mean *make certain* and *insure* to mean *cover with insurance.*"

- **Hyphenation.** In addition to cross-referencing your preferred style guide or dictionary, list frequently used terms that require or don't require hyphenation: In your documents, is it *day-trading* or *daytrading*? Fiberoptic or *fiber-optic*?

CASE STUDY: USING A MACRO TO CHECK FOR BIASED LANGUAGE

Problem: A large national organization needed a way for staff members to check for and correct unintentionally biased language in a wide variety of business communications. Although memos could inform staff of the need to check for potentially biased terms, it was difficult for everyone to remember and correct the hundreds of terms that could be problems. The company decided to have a programmer create a macro that would search for and replace problem words. One challenge they faced was that although some problem words could be corrected through a simple search and replace—for example, changing *deaf and dumb* to *deaf or hearing impaired*—others required more intervention. For example, changing *handicapped* to *person with a disability* in the sentence *This is an excellent guide for handicapped people* required the writer to manually delete *people*.

Solution: The programmer created a macro that functioned like a spelling checker, but instead of checking spelling, it checked for potentially biased wording. The writer clicked a button to activate the macro, and the macro searched the document for problem words from a predetermined list. When the macro located a problem word, the writer could read the surrounding text, choose from suggested replacement words shown by the macro, and make the appropriate substitution. If necessary, the writer could pause the macro and edit the text extensively, then restart the macro to continue checking the document. The macro was distributed on diskettes and was easy to install on any computer that had the required word processing program (Microsoft Word).

- **Grammar.** This is one topic that can quickly get out of hand. You aren't writing a grammar text, but a few cogent examples can save people a lot of trouble. One way to home in on the problems that trip people up is to ask your editorial staff what grammatical errors they keep having to correct in the work they review. Once you post a brief grammar section on your intranet, it will grow as people submit questions—especially if you do informal surveys. And you can encourage your writers to consult this section and ask questions, too.

  Perhaps the most vexing question in grammar is that of subject/verb agreement for collective nouns: *A small percentage/a number of people/the majority* **is** or **are**? Provide a rule and several examples to illustrate it.

- **Capitalization.** Stick to a couple of simple rules (e.g., "Capitalize job titles before, but not after, a person's name: *Chief Executive Officer Smith,* but *Mary Smith, chief executive officer.*" "Do not capitalize the words *table* and *figure* in text: *See figure 2.*" "List specific terms that should always be capitalized, and terms with unique capitalization: *1stUp.com.*"

- **Punctuation.** The main discretionary items are the use of serial commas *(Tom, Dick and Harry* versus *Tom, Dick, and Harry),* capitalization after colons, whether to place end punctuation inside or outside an end quotation mark, and punctuation with lists. But the punctuation section, like the grammar section, may grow as people submit questions about what they find confusing. Always ask for example sentences that show the problem in context.

- **Numbers.** You will need to go into some detail here because there's a lot of latitude in when to use figures versus numbers. You may simply adapt one of the published styles, but provide examples drawn specifically from your organization's topics. This section can cover the use of numbers in text and in lists, tables, and sidebars, as well as basic rules: "Round up to whole numbers in comparisons" and "Don't start a sentence with a figure."

CASE STUDY: USING A MACRO TO CONVERT TABLES

Problem: A publishing company produced more than 60 newsletters a week and needed a way to quickly strip the electronic files of extraneous layout codes and convert the files to multiple coded text files (including HTML) that could be sent electronically to news wires. Because of the immediate turnaround, the accuracy required by the news wires, and high error rate of the manual process, the company needed a faster, more economical, more reliable way to process the files. A programmer created an all-encompassing macro that would turn these newsletters into working coded text files. One challenge to be faced was converting tables. The company had many different types of table formats, but on its own, the macro could recognize only one type of table. For example, one table might have headings down each row that should instead be at the top of each column, while another table might be the opposite, and still another might be one large table broken into smaller tables. How can the macro tell the difference?

Solution: It can't—but the editor can. The programmer devised a macro to act as an interactive menu. The editor could then highlight a table and click a key to activate the macro. A menu then prompted the editor with a list of possible table types. The editor clicked a button for the desired type, and the macro converted the table appropriately. If this wasn't what the editor wanted after all, the original table was retrieved, and the editor was again presented with the menu of options. This macro was part of a larger text conversion macro and was distributed through the company's network on the editors' computers, which had the required word processing program (Microsoft Word).

- **References.** This section can easily become a book in itself, but if you publish references, your house style should give at least a few typical examples (e.g., book, journal, Web site) along with a cross-reference to your default style. Include in-text and sources-used citation styles and a footnote style.

- **Format.** How many types of documents do your users produce? You may want to provide electronic templates for the most common. For a large organization with many departments, each of which has been producing documents with its own format, the creation of a house style can be the opportunity to achieve consistency. But beware: Format can be a contentious topic if you don't have corporate style or branding guidelines to follow.

Format includes both editorial and design considerations. Here are some of them:

- Type choices: font defaults and permissible combinations
- Levels of headings
- Running headers and footers
- Logos
- List style
- Table and figure style and file format
- Address information (telephone numbers, state abbreviations)
- Fax and report cover sheets
- E-mail signatures
- Colors for branding pieces (letterhead, report binding, leave-behind fact sheets and brochures, press kit folders), presentations, and Web sites

Making Specific Editorial Choices

As long as you don't give in to the temptations of whimsical laissez-faire, you can designate as "officially correct" the editorial choices that seem right to your eye and ear. Wherever alternatives exist, you'll need to provide users of your guide with examples that offer guidance in context—that means in the context of the sorts of documents they will be writing and editing.

One way to be sure your guide is relevant to your style issues is to test it before distributing it to everyone or going live online. Have a few editors work through it critically, and ask a few editors, writers, and managers to try using it for a month or so. Ask them to note what helped and what didn't. As they encounter omissions, mistakes, and incomplete cross-references to related material, your beta-testers will give you a chance to strengthen the usefulness and accuracy of the guide. And you're setting a precedent for updating the guide; you'll need all the feedback you can get from now on.

Here's a suggested list of priorities once you've got agreement that a proprietary guide is called for and you're the lucky one designated to create it:

- Set up a committee of the people most qualified to identify the points of confusion, research the options, and make editorial decisions. "Most qualified" means by dint of wide, not necessarily long, experience as a hands-on writer or editor, and a flexible outlook that will allow them to make pragmatic decisions instead of going ballistic because something "isn't in Webster's."

- Ask everyone who produces documents to meet for 30 minutes to kick off the planning stage. Ask for a few volunteers right then and there to help prepare and send out a company-wide e-mail survey of "things that drive you nuts or worry you most," compile results, and set the stage for recommendations.

- Target the most frequently cited style problems for immediate remedy and circulate draft guidelines for review by everyone who prepares documents, not just management. (Remember to include the layout and production specialists.) Ask for comments.

- Collate comments and distribute "final" style decisions and some examples by e-mail or paper memo. Congratulations! You have begun preparing a guide by treating the most irritating editorial hotspots.

- Working with management and your art director or senior design staff, start a second stage of formalizing editorial and graphic standards that everyone who prepares documents should be aware of.
- Make templates of letterhead logos, addresses, taglines, trademarks, and document formats (memo, fax cover sheets, letter bids, formal proposals) available to all staff on the intranet.
- Ask whoever performs copyediting and quality control reviews to make note of or keep copies of the style problems they see most often. These are likely to be the areas people need most education about—because they don't know when to spell out numbers, they have the rules wrong ("There's no difference between it's and its"), or they don't know that there *is* a style decision to be made about capitalization of the first word of a clause after a colon.
- Compile these refinements and nittygritty examples and distribute them to everyone for comment. Ask people to look for examples of when the rules don't work or just to tell you whether they "like" the rules you propose. You may not change the rules, but giving people a chance to be grouchy now can pay off later—what you really never want to hear later on is, "Why should I do it that way? 'Our' style, huh? Nobody asked *me* about it."
- Start planning the best way to get a copy of the complete set of guidelines on every desk or computer, and set a schedule for regular review and updates. Keep asking people to give you examples of problems from their work that need resolution so style decisions can leave the world of anecdote and personal preference and become part of the guide.
- Thank everyone, personally and publicly, who has contributed to the guide.
- Don't let down your guard. A style guide is always a work in progress and will need revision as often as your organization repositions itself, adds jargon for new processes, and redesigns itself.
- Make sure to tell people whenever you depart from your designated main style guide and main dictionary—and make sure

people have copies of your base references within easy reach for filling out the gaps in your guidelines as you continue to build them. Better to have people making somewhat consistent decisions in the meantime about, say, when to hyphenate or close up new compounds, even if you are still in the process of making changes to be more consistent with other in-house style decisions.

Proprietary Styles at Work
To get a sense of the range of editorial and graphics details that a style may encompass, it's revealing to see style decisions at work (or not working so well) in context. Styles carry implications, and making one good decision often calls for another. Here are a few examples:

- The Copyright Clearance Center (CCC) calls its online stock image service M*i*ra (Media Image Resource Alliance). The CCC style guide, if there is one, should point out the correct rendering of the name's special graphic treatment in all press releases, correspondence, and brochures in order to reinforce it as a trade name. Assuredly the rest of the world is going to try to avoid typing that slowdown of a mid-italic. And in fact, the e-mail address (mira@mira.com) can't, and the Web site text doesn't, employ the italic.

- The Web site of AARP (formerly the American Association of Retired People) doesn't bill itself as a site, but as the AARP webplace. Newcomers are invited in some printed material to *visit our webplace at…*, but on the site people are also invited to find out *what's new on Webplace* by subscribing to a *Webletter*. We can also *EXPLORE WEBPLACE.* That much variation in capitalization is in need of standardization. But on Web sites, especially, it's standard operating procedure to present trade names in an array of styles; in any medium, it's asking readers to track trivial distinctions.

- The managing editor of an IBM magazine for AS/400 computer users meets every six months with her staff to discuss the acronyms most frequently used in the publication. They try to decide which ones are so commonly used in the industry that they don't need to be written out, even at first reference. Some of the acronyms on their list are API, CPU, FTP, LAN, SQL, IT, and ISP. They still worry that not all readers know what those terms mean, yet some are used over and over in the issue and can't be repeatedly redefined. This is an ongoing struggle for the magazine's staff.

- The Corporation for National Service's editorial guidelines begin with an excellent preface that gives users an overview of the goals for Corporation publications; nothing is taken for granted. Users are reminded that "For any type of publication...the language, tone, length, and appearance should reflect your intended audience and purpose," and the possible audiences are explicitly listed: potential national service participants, current national service program directors or members, older individuals, college students, or individuals in the private sector. Writers are cautioned that some readers may have never heard of the Corporation, and also to consider what a piece is trying to accomplish. Specific advice is highlighted:

  > Don't overuse abbreviations, and do not use acronyms unless they are terms you want your readers to know. For example, *AmeriCorps*NCCC, AmeriCorps*VISTA,* and *RSVP* are terms we want the public to be familiar with, while *FGP* (the Foster Grandparent Program), *SCP* (the Senior Companion Program), and *NSSC* (the National Senior Service Corps) are not.

- *The Editorial Eye* newsletter has a stylesheet adapted from several styles, in a mix both conservative and forward-looking. State abbreviations are postal-style abbreviations (all caps). Many terms that *Merriam-Webster's Collegiate Dictionary,* tenth edition, and other styles show as still two words are rendered as a solid compound to reflect today's wide use (*badmouth, inhouse, onsite, freelancer, copyedit*). Yet the newsletter holds the line against contemporary gaffes such as hyphenating *-ly* adverbs (as in *The barely-tasted cookie fell to the floor*) and using a colon with a phrase that leads in to a bulleted list and a sentence that is completed by each of the items in the list (as in *The three main uses for the colon are*).

An anecdote may put things in perspective and help you keep a sense of humor (you'll need it) about this style guide business. Phil Smith III, an external reviewer of this book, passed along a story from his father: "The government translation agency that I used to work for put out a number of bulletins concerning terms to be used in translating. For example, how do you translate *oblast* from Russian? Is it a country, a district, or what? For users to understand what is going on—when Gorbachev has been promoted to head honcho of Zhivkovast, just what office does he now hold?—we had to agree on terms for translation. The publication that was produced to make these editorial decisions universal was called *No Uncertain Terms*."

That's really what you're trying to do: help editors and writers and designers lose their uncertainty so your readers and users won't be bothered by unnerving little discrepancies and holes.

MAKING INTRANET CONTENT EASY TO USE

Which is more frustrating to people in search of help with style? Having to flip through pages of badly organized and cryptically indexed rules, or having to scroll through long screenfuls of unsearchable items with many irrelevant links to get to the point? It's a wash. Very few people read style guides—in any medium—for fun.

What you learn here about online navigation has implications for print, too. It's hard to beat great categorical organization and a great index, and it's hard to beat an intelligent keyword search capability based on HTML with supplemental material in a PDF to show examples of graphics (forms, list styles, address formats, acceptable color and fonts, logos, and the like). In both mediums, these are the most important things to build in:

- Recognizable lists or categories of problems and choices.
- Logically cross-referenced answers that can be located quickly.
- Examples that make distinctions, exceptions, and applications clear.

The Case for an Online Guide

If users can zero in on the questions they need answered by using keyword searches or creating bookmarks to the topics that perennially trouble them, they'll be less likely to "wing it" or "go by ear," two great ways to sabotage a style guide. And what an online style can do that even the best-indexed paper version just plain can't is offer rapid and comprehensive searching and cross-referencing.

As the *Yale Style Guide* puts it, an intranet should let you "get in, get what you want, and move on." The section on intranet site design will be useful for any editor who is not familiar with the principles of online usability and navigation; here's the gist: "Successful intranet sites assemble *useful* information, organize it into *logical* systems, and deliver the information in an *efficient* manner" (emphasis ours). For more, go to http://info.med.yale.edu/caim/manual/sites/intranet_design.html.

Time is money, especially when employees are trying to find an answer so they can get back to work. The single most important aspect to keep in mind when creating content for an online guide is the user. Providing navigation from place to place and interactive links within pages will help users get a sense of "you are here," without which "over there" may seem irrelevant instead of enriching.

INTELLIGENT LINKS

One of your biggest challenges in writing an online style guide will be placing intelligent links. The trick is to embed them in the natural conversational flow of a statement—The *Yale Style Guide* calls it *parenthetical* placement—so the link makes logical sense without interrupting. Avoid saying, *Click here for more examples of dangling modifiers.* Avoid giving the imperative *Click,* period. Why? Such links are distracting—when someone is trying to find information, it's annoying to be commanded to go somewhere else for it. Group all minor links and footnotes at the bottom of a section, not in the main text. Phrase important links in this manner: *Watch out for misplaced modifiers—dangling, wandering,* and *squinting—that create unwittingly humorous descriptions.*

One of the best resources for intelligent online design is still the *Guide to Web Style* by Rick Levine, published by Sun Microsystems.

A good overview of intranets in general—how and why to set them up and what the benefits, obstacles, and system requirements are—can be found at www.intranetroadmap.com, where OpenConsult, Inc., an intranet service provider, has provided an Intranet Road Map. A statistic quoted on the site: Eighteen percent of corporate printed material becomes outdated after 30 days. Online material can stay current—certainly a benefit for a corporate style guide. Two-thirds of Fortune 1000 companies had an intranet as of July 1996, according to the site, but it's a guess how many contained a company-wide style guide. Style guide didn't make it among the listings of possible uses including calendars, groupware applications for collaboration with consultants and partners, weather and traffic links, internal departmental information, telephone directories, user documentation, and so forth.

The truth is, online style guides don't rank high in the mind of anyone who isn't engaged in quality control of publications. But for those of us charged with saving organizational face and trying hard not to reinvent the wheel each time we edit a document, they're a natural, and they're essential. In fact, many corporation intranet sites are maintained by only one dedicated staff person. The basic requirements for a simple intranet site are

- A TCP/IP protocol with enough bandwidth to handle all types of multimedia.
- Firewalls to keep hackers and crackers out and to protect remote location rights.
- A Web HTTP server (an intranet doesn't have to be hooked up to the Internet).
- For employees, a Web browser.
- For authors, authoring and development tools.
- For all, access to search tools, bookmarks, printing, and graphics support for reading PDF files.

Tech writers might need a style guide as complex as a self-extracting multimedia presentation with streaming video or as simple as the Babel Glossary of Computer Related Abbreviations and Acronyms at www.geocities.com/ikind_babel/babel/babel.html, which also includes a list of over 200 recognized country domain names for Internet addresses.

The Babel glossary can be kept on a hard drive or a floppy disk (downloading constitutes an agreement to outlined terms) or printed out. It is revised three times a year by its creators, Irving and Richard Kind, and it can be searched five different ways—and all of this information and more is explained to users on the opening page. The search functions include four that any style guide could also use: alphabetical letter-group buttons for entries, the Page Down key, the vertical scrollbar, and the browser Find command. For a showcase of Web sites that are models of how posting effective signs can help the reader access information, visit www.ewriteonline.com/showcase/index.html.

As the *Yale Style Guide* says, "Graphic user interfaces were designed to give people direct control over their personal computers….The goal is to provide for the needs of all your potential users, adapting Web technology to their expectations, and never requiring the readers to simply conform to an interface that puts unnecessary obstacles in their paths."

Here's a summary of what an efficient intranet site also tries to do to speed users through the search and skim processes basic to consulting a style guide:

- Provide menus; tables of contents; button bars that let users go back, forward, or return to the opening page or a related menu page; and short summaries of what can be found on other pages. In short, give people who search differently a choice of paths leading to the same information.

- Highlight keywords, write meaningful heads and subheads, and incorporate a limited number of relevant links. (Too many links discourage users. Irrelevant links infuriate users.)

- Be concise, organize information into short paragraphs, and don't use a line length the entire width of the screen. (On most monitors, there is usually a little bit of the "page" that a user can't see. Make sure that vital information is placed in the upper half of the screen.)

- Use hypertext links to accommodate the needs of many different users working on different types of publications, for different media. Links will lead users to relevant pages for their particular project and allow them to bypass the rest, avoiding a sense of "too much information."

If these quite basic ideas are new conceptual territory for you, it's time to have a meeting with your favorite resident programmer or webmaster to explain the scope and use of the guide you have in mind and ask for help. Before you begin compiling content, think of how it must be organized for searching and linking, and learn enough about screen design to avoid parking a print piece online that fails to take advantage of interactivity and navigation tools.

Design and Accessibility Considerations
According to the *Yale Style Guide,* readers see pages and screens first as a blur of shape and text blocks and color. Then they begin to pick out pieces of information. Type and illustration can play a part in helping or can clutter the landscape with even more cues that are hard for readers to process. So hierarchy still matters very much online, but is not necessarily achieved by using the same physical cues a printed reference provides.

The purpose of online design is to create a consistent, recognizable, simple plan whereby important elements look the strongest and content categories are logically organized and predictably flagged. (Don't make users guess when an underline is a live link and when it's an underline.)

We ask our readers to trust us when they see a link to another screenful of information. (How much text? Will it be relevant, or just a silly, slow-loading graphic?) They have no assurance that they'll find a useful design template or example sentences at the other end.

Functional cues must give users a way to understand how information is organized and how much attention they should give to certain parts of it in order to get to what they need. The *Yale Style Guide* unequivocally states,

> Editorial landmarks like titles and headers are the fundamental human interface issue in Web pages, just as they are in any print publication. A consistent approach to titling, headlines, and subheads in your document will aid your readers in navigating through a complex set of Web pages.

> **REALITY: THEY'RE GOING TO PRINT OUT PAGES**
> Face it: Most people want a cheatsheet to the problems they encounter most often. Online pages containing a lot of text should be designed for printing (i.e., not extra-wide), because that's what people do with important online information. They want to have it at hand to absorb and refer to. You don't want users to lose a couple of words off the right margin of the printed page. Recommended online page layout "safe area" dimensions for printing (from the *Yale Style Guide*) for a 640- by 480-pixel screen are as follows: maximum width, 535 pixels; maximum height, 295 pixels.

Taking care to organize a useful online reference really requires creating a style guide for the style guide! But that's not a problem—it's a chance to reinforce the guidelines by showing them at work. Decisions about using text styles consistently and emphatically in an intranet style guide might result in the following set of specs—which, ideally, would reflect the recommendations in the corporate style guide:

Headline style
Bold, capitalize initial letters—for document titles, other Web sites, proper names, product names, trade names

Downstyle
Bold, capitalize first word only—for subheads, references to other headings within the style manual, figure titles, lists

And, adapted from the Yale guide, a checklist for enhancing readability:

- Direct the reader's eye toward important information right away with strong type.
- Use subtle pastel shades of colors for background.
- Avoid bold, saturated colors except for accent or infrequent spots of emphasis.
- Make sure type contrasts sharply against any background color—black type is best; with dark backgrounds, a lot of reversed white type can be hard to read.
- Beware of graphic embellishments you might use reflexively on paper—horizontal rules, bullets, icons, large display type sizes. They may look grotesque on your reader's browser.

MAKING IT WORK

Successful corporate style guides have these characteristics in common:

- They are introduced to the organization with enough fanfare that people realize that they are to be taken seriously;
- They are updated frequently enough that users regard the advice as credible and reliable; and
- They are presented and viewed as helpful tools, not as weapons to use against those who don't conform.

The Rollout
The method you choose to introduce the style guide to your organization can significantly affect its success. You want your colleagues to regard the style guide as an essential tool, not as just another corporate manual that sits on the shelf (or on the intranet, as the case may be). When the guide is finally ready to be released and you are planning its distribution, consider the following approaches:

- Get buy-in from the top. Ask the CEO to sign a letter that discusses the organization's commitment to quality and the style guide's role in ensuring that the organization's written products—whether printed or electronic—represent it well. The overall message should be that everyone shares the responsibility for upholding the organization's quality standards.

- Invite the users to a kickoff party. Give a presentation in which you explain how the guide is organized and how it should be used, but add an element of fun as well. One organization launched its guide with a contest, presented gameshow-style, in which teams of participants scored points by answering questions about the company's editorial and design standards.

- Make it easy for people to ask questions. Though your ultimate goal is for people to use the guide rather than picking up the phone, give them a human point of contact (not just an e-mail box) to help them familiarize themselves with the details. If people are consistently having trouble with a particular section, you'll know what to focus on when the guide is updated.

Updating the Guide
The process of updating a style guide should be determined by the needs of the organization and the rate at which the information will go out of date. Realistically, the guide will probably be out of date in some respect the minute it is released. The question is, how often do *significant* changes occur that affect the usefulness and credibility of the guide? For a technology-oriented company, that might be once a month, because product names and version numbers are key elements of corporate documentation. For an association, updates might be needed only once a year. The job

of maintaining the style guide can take on a life of its own if its contents are not monitored by someone with a sense of proportion about the importance of changes.

Updating the style guide gives an organization the opportunity to ask itself the all-important question "Why do we do it this way?" The very process of documenting decisions can highlight standards that have become obsolete. Sometimes specifications can be traced to old technology (maybe the requirement to produce all presentations in Helvetica started because that was the only typeface available on the network printer in 1992). Make the most of the update by questioning old practices and changing those that no longer make sense.

Here is one approach for managing the update process:

- Once the style guide has been distributed, designate someone to be the collection point for questions. Include a feedback mechanism (a simple, short printed or electronic form) with the guide to make it easy for people to offer comments.

- Decide whether you will update the guide frequently to incorporate minor revisions (and call it Version 1.1) or less frequently to incorporate major changes (and call it Version 2.0).

- Define the review process. Minor updates might require the eyes of only a few reviewers; major updates should be routed to the entire development committee, and probably to a sampling of users.

- For minor updates, indicate what changed (either with a cover memo or by inserting change bars in the margins) and redistribute the guide. For major updates, consider convening the users for another presentation of the changes. Major changes should include some discussion—whether at the presentation or in the guide itself—of the rationale behind the changes. And if there is no rationale, explain why an exception to conventional wisdom is being made. Users will be far more likely to change their ways if they understand why they are being asked to take a different approach.

A WEB SITE NEEDS A STYLE GUIDE, TOO

Web sites need a consistent style to guide their content providers and maintenance staff. A good dynamic site tends to grow quickly as new kinds and levels of information about products, services, references, and people are added. New content shouldn't be added without regard for consistent style across pages, which will also make updating the site easier. HTML formatting tags can be done several ways, as Rick Cook explained in his April 1997 "Best Practices" online column in Netscape World (now called Netscape Enterprise Developer):

> For example, you can indicate emphasis with <STRONG> or <EM> or <B> or in other ways. In fact, you can do things several ways and have them look the same—sometimes. On some browsers <EM> is interpreted as boldface, the same as <B>. On other browsers, <STRONG> might produce bold text and <EM> might italicize the text—which <B> will never do. So you have two different methods of emphasizing something in the same document doing two different jobs. Confusing? Yep...Sometimes the results of these inconsistencies are worse than confusing for the [site] visitor. Sometimes they are hair-tearing, time-eating nuisances for the people trying to maintain the site.

As Web sites grow, integration of the distinctive elements that tell visitors they are visiting *your* site becomes a primary style task. Controls should work the same on every page; backgrounds, section colors, the name of your organization, addresses, logo treatment, trademarks symbols, copyright statements—all of it should be coordinated. Make sure your guide advises against using what Cook calls "browser-specific gimmicks" that detract from the information mission. Each Web page should be proofread before it goes to the site, and the chain-of-approval process should be clear. If more than one person works on the site, designate who can authorize exceptions to the style and in what cases. For a list of Cook's Best Practices columns, visit www.netscapeworld.com/netscapeworld/common/nw.backissues.column.html#bestpract.

The Reality of Guidelines

There is one final bit of advice that should be included in any discussion of guidelines: No matter how careful you are to make your style guide easy to use and understand, there will always be a few people who simply choose to ignore it. (These are the same people who don't read their human resources policy manual and wonder why they are suddenly out of leave.) Don't think of these people as your target audience; you'll just end up banging your head against the wall of your cubicle. The real target audience for the guide is the people who—though they may not be documentation specialists by training—want their written products to be clear to the intended audience and produced in an efficient manner. We'd be willing to bet that most of the people in your organization fall into this category.

Annotated Bibliography

ELECTRONIC SOURCES

American Dialect Society (ADS) Web site, www.americandialect.org. ADS bills itself as "the only scholarly association dedicated to the study of the English language in North America." The site features rare finds such as the American Dialect Society's official list of "words of the year, decade, century, and millennium."

Chicago Manual of Style online FAQ, by the manuscript editing department at the University of Chicago, www.press.uchicago.edu/Misc/Chicago/cmosfaq.html. Features commentary on issues—such as new media terminology—that have arisen since the last major revision (in 1994) of the *Chicago Manual of Style*.

The Maven's Word of the Day, by Random House, www.randomhouse.com/wotd/. This site defines and traces the origins of new words, many of which are tech-related.

Dale Dougherty. "Don't Forget To Write." *Web Review*. http://webreview.com/97/10/10/imho/index.html (Jan. 21, 2000).

"Interactive Publishing." Electric Pages. http://in3.org/maxims/max_2.htm (Jan. 21, 2000).

Nick F. Nichols. "20 Web Site No-No's!" www.internetday.com/archives/061698.html (Jan. 21, 2000).

Steve Outing. "Hey, Online Writers: It's Time to Focus on the Writing." www.mediainfo.com/ephome/news/newshtm/stop/st051198.html (Jan. 21, 2000).

Jack Powers. "Writing for the Web, Part I." www.electric-pages.com/articles/wftw1.htm (Jan. 21, 2000).

___."Writing for the Web, Part II." www.electric-pages.com/articles/wftw2.htm (Jan. 21, 2000).

Nancy DuVergne Smith. "Writing for the Web." http://members.aol.com/nancyds/webwrite.html (Jan. 21, 2000).

Brad Templeton. "10 Big Myths about copyright explained." www.templetons.com (Feb. 2, 2000).

Carl Thress. "Writing for the World Wide Web." www.thescratchingpost.com/wordsmithshop/writing/writinga.html (Jan. 21, 2000).

Wired Style online, hotwired.lycos.com/hardwired/wiredstyle/toc/index.html. This site featuring the first edition of *Wired Style* has not yet been updated to reflect content or even the existence of the revised and updated edition. Nonetheless, it contains a rich assortment of new media words and references for more information.

Word Spy, by Paul McFedries, www.logophilia.com/WordSpy/index.html#subscription. By its own description, "This Web site and its associated mailing list are devoted to recently coined words, existing words that have enjoyed a recent renaissance, and older words that are now being used in new ways." The daily word mailings are refreshing reading, and always contain recent citations from the popular press to show usage.

World Wide Words, by Michael Quinion, www.quinion.com/words. Quinion's site features a British perspective on the evolving language. His Web site and (always welcomed) monthly e-mail newsletter explore the quirks, curiosities, and evolution of the English language.

GLOSSARIES, THESAURUSES, AND ACRONYM LISTS ONLINE

Use these sources freely, but wisely. As we delved deeply into research on words for the glossary, we found that most of the technical glossaries listed below differed not only in spelling and punctuation, but at times were also at odds in their definitions of terms. The consensus approach is highly recommended. When researching a word, search several of these sites and look for points of intersection. OneLook Dictionaries (see below) is an excellent place to start, as it searches hundreds of glossaries in one pass, including several of those listed here.

Acronym Finder, by Mountain Data Systems, www.acronymfinder.com. A comprehensive database of English-language acronyms (not limited to tech terms) with sophisticated search capabilities.

Babel: A Glossary of Computer Oriented Abbreviations and Acronyms (version 00A), by Irving & Richard Kind, www.geocities.com/ikind_babel/babel/babel.html. This site features extensive lists of technology-oriented acronyms.

CNET Glossary, by CNET, Inc., http://coverage.cnet.com/Resources/Info/Glossary/index.html. CNET is a major Web-based source of information on computers, the Internet, and digital technologies. This is a nice, tidy glossary.

Dictionary.com, an online English Dictionary and directory of Internet reference sites, by Lexico LLC, www.dictionary.com. One of the better dictionary sites on the Web. Dictionary.com enables word searches in references such as *Webster's Revised Unabridged Dictionary, The American Heritage Dictionary of the English Language, Third Edition,* Roget's Thesaurus, and Princeton University's WordNet. Another section of the site features an online translator that translates text and Web pages from English into several foreign languages.

Free On-Line Dictionary of Computing, editor Denis Howe, http://foldoc.doc.ic.ac.uk/foldoc/index.html/. This extensive dictionary contains 13,000 computer-related terms. It is also available on CD (see http://burks.brighton.ac.uk/ for more information).

The Jargon File online (4.1.2 version), by Eric S. Raymond, www.tuxedo.org/~esr/jargon/. This site deserves pages of praise for its thorough documentation of both computer geek culture and the language it has engendered. It is the living, breathing (regularly updated) source from whence the *New Hacker's Dictionary* (MIT Press, 1996, ISBN 0-262-68092-0) sprang.

Onelook Dictionaries, by Bob Ware, www.onelook.com. This site enables rapid searches of *hundreds* of online dictionaries, glossaries, and acronym lists. This site and dictionary.com are our top two picks.

NetLingo: The Internet Language Dictionary, by Vincent James and Erin Jansen, www.netlingo.com.

SpellWeb, by Clear Ink, www.spellweb.com. This site cleverly automates a technique for looking up words on the Web that has become commonplace. It features a form for entering a word, spelled two different ways (such as *antialiasing* and *anti-aliasing*).

SpellWeb submits the two different spellings to a major search engine such as WebCrawler, and returns results on how many times each spelling of the word was used on the Web. While this technique still doesn't ensure that the word is spelled correctly, it is a reassuring populist opinion poll.

Thesaurus.com, an online version of *Roget's Thesaurus of English Words and Phrases,* by Lexico LLC, www.thesaurus.com.

Webopedia, by internet.com, www.pcwebopedia.com. This was one of our favorites, a very detailed, extensive technical glossary, though as a generic term, webopedia ranks right up there with webrary as sadly lacking euphony.

E-MAIL DISCUSSION LISTS

The American Dialect Society has an e-mail discussion list, accessible at www.americandialect.org/adsl.shtml. On this list, members of the society and other word lovers discuss origins of words to an exquisite depth of detail.

COPYEDITING-L. Perhaps the definitive electronic mailing list for copyeditors. Archives and list sign-up are available at http://listserv.indiana.edu/archives/copyediting-l.html.

Online-Writing List (OWL), operated by Content Exchange LLC. This is an e-mail discussion list for writers, editors, and producers of online content. Access the list at http://www.content-exchange.com/cx/html/owl.htm. The list delves frequently into issues of online copyrights, contracts, and rates for services.

PRINTED BOOKS

Computing Dictionary: The Illustrated Book of Terms and Technologies, fourth edition. Lincoln, NE: Sandhills Publishing, 1997.

Dodd, Janet S., ed. *The ACS Style Guide,* second edition. Washington, DC: American Chemical Society, 1997.

Freedman, Alan. *The Computer Glossary: The Complete Illustrated Dictionary,* eighth edition New York, NY: American Management Association, 1998.

Hale, Constance, and Jessie Scanlon. *Wired Style: Principles of English Usage in the Digital Age.* New York, NY: Random House, Inc., 1996.

McGuire, Mary, et al. *The Internet Handbook for Writers, Researchers, and Journalists.* Toronto, Canada: Trifolium Books, 1997.

Merriam-Webster's Collegiate Dictionary, tenth edition. Springfield, MA: Merriam-Webster, 1993.

Microsoft Manual of Style for Technical Publications, second edition. Redmond, WA: Microsoft Press, 1998.

Microsoft Press Computer Dictionary, third edition. Redmond, WA: Microsoft Press, 1998.

Pavlicin, Karen, and Christy Lyon. *Online Style Guide: Terms, Usage, and Tips.* Saint Paul, MN: Elva Resa Publishing, 1998, www.elvaresa.com.

Raymond, Eric S. *The New Hacker's Dictionary,* third edition. Cambridge, MA: MIT Press, 1998.

Siegal, Allan M., and William G. Connolly. *The New York Times Manual of Style and Usage.* New York, NY: Times Books, 1999.

Read Me First! A Style Guide for the Computer Industry. Mountain View, CA: Sun Technical Publications/Prentice Hall, 1996.

Walker, Janice R., and Todd Taylor. *The Columbia Guide to Online Style.* New York, NY: Columbia University Press, 1998.